Echoes of War

The Story of H$_2$S Radar

Echoes of War

The Story of H$_2$S Radar

by

Bernard Lovell

Adam Hilger
Bristol, Philadelphia and New York

British Library Cataloguing in Publication Data
Lovell, Bernard
  Echoes of war: The story of H$_2$S radar.
  I. Title
  621.384809

ISBN 0-85274-317-3

Library of Congress Cataloging-in-Publication Data are available

Published under the Adam Hilger imprint by IOP Publishing Ltd
Techno House, Redcliffe Way, Bristol BS1 6NX, England
335 East 45th Street, New York, NY 10017-3483. USA

US Editorial Office: 1411 Walnut Street, Philadelphia, PA 19102

Typeset by MCS Ltd
Printed in Great Britain by J W Arrowsmith Ltd, Bristol

Contents

	Preface	vii
	Author's Note	x
	Chronology of Radar Developments	xi
1	August 1939	1
2	Scone Airport	10
3	St Athan	20
4	Worth Matravers	27
5	Leeson House	45
6	AIS—the First Centimetre AI (AI Mark VII/Mark VIII)	55
7	Lock–Follow AI (AIF/AISF/Mark IX AI)	69
8	H₂S—the Background	85
9	The Birth of H₂S—1 November 1941	91
10	Halifax V9977	99
11	Life in Swanage 1940–42 (by Joyce Lovell)	110
12	The Last Days in Swanage and the Food Queues of Great Malvern	119
13	The Crash of the Halifax Bomber	126
14	The Meeting with the Prime Minister	132
15	Bennett and Renwick	137
16	Autumn 1942	142

17 January–February 1943   151

18 Centimetre ASV and the U-Boats   155

19 The Impact of the German Naxos on Centimetre ASV   165

20 H₂S on Tank Landing Craft   171

21 The Summer of 1943—Destruction of Hamburg   173

22 H₂S on 3 Centimetres (X-band)—the Attacks on Berlin and Leipzig   180

23 H₂S and the American 8th Bomber Command   193

24 The Problems with H₂S in Bomber Command   197

25 Fishpond   206

26 Conflict with Bomber Command   211

27 The New Versions of H₂S   217

28 July 1944   229

29 Naxos and H₂S   233

30 D-Day, H₂S and the Army   238

31 The U-Boat Schnorkel   246

32 The Last Months of the War and the Post-war Phase   251

33 Envoi—1991   261

Appendix 1: Note on Minelaying by H₂S   263

Appendix 2: Glossary and Abbreviations   265

Appendix 3: Principal Staff and Command Appointments   273

Appendix 4: Summary of H₂S Systems   275

Index   277

Preface

In August 1939, like many other young people pursuing academic research in the universities, I was suddenly plunged into an entirely different kind of research and development. I listened to Prime Minister Chamberlain's broadcast on Sunday morning, 3 September 1939, announcing that we were at war with Germany, in the operations room of one of the giant coastal defence radar stations. The radar echoes I saw on that morning were not from enemy aircraft. The search for their origin six years later led me back to the University of Manchester, to Jodrell Bank, and to another kind of research. I have described the consequences in *Astronomer by Chance*.[1] This present book concerns my life and work on radar during the years of World War II.

E G (Taffy) Bowen was one of the few pioneers in the development of British radar in the pre-war years. In *Radar Days*[2] he has described those pioneering days that led to the radar coastal defence network which was to be such a vital factor in the defeat of the German Luftwaffe during the Battle of Britain in 1940. Before the war began Bowen had developed the first radar to be carried in our night fighters. This airborne development was initiated because it was believed, correctly as it turned out, that the German Luftwaffe would be defeated by day and would then turn to bombing by night. It was to Bowen's airborne group that I was sent in September 1939 and the next seven chapters of this book describe my work on this airborne interception radar (AI). However, as it transpired, this early work soon led to the development of airborne systems working on much shorter wavelengths than the 1½ metre wavelength of Bowen's original system. In the summer months of 1940 I was involved in the initial revolutionary application of the magnetron to airborne radar. This means of generating considerable amounts of power on a wavelength of ten centimetres and less was discovered by Randall and Boot in the University of Birmingham in February 1940, and the magnetron soon revolutionised other radar developments as well as the AI system.

In the autumn of 1941 the failure of Bomber Command to locate its targets in Germany came to light and in the last days of that year I was ordered to abandon my work on centimetre AI systems in order to form a new group to develop a blind bombing system to be carried in Bomber Command aircraft. By that time Bowen had gone to America to assist with the US radar development and I was no longer working with him. After the publication of his book in 1987 he urged me to carry on the story by describing the development of this blind bombing device known as H_2S and its conversion to assist Coastal Command in the detection of U-boats. The development of the H_2S blind bombing system was carried through under extraordinary and tragic circumstances and was first used operationally by the Pathfinder Force in the attack on Hamburg during the night of 30–31 January 1943. Two months later a few of these equipments, modified and fitted in Coastal Command aircraft operating over the Bay of Biscay had a dramatic tactical effect on the air war against the U-boats.

In later chapters of this book I describe the development of improved forms of the blind bombing H_2S, and the difficulties and antagonisms that faced the system. The Prime Minister had demanded the highest priority for the development and production of H_2S. This caused many problems but also meant that for some time the electronic units made for H_2S became the major source of centimetre radar. As well as the adaptation for use as ASV (Air to Surface Vessel) by Coastal Command a modification was fitted on tank landing craft and used during the invasion of Sicily. Although at first the Americans were sceptical about the value of H_2S, when the US 8th Bomber Command began operating over the cloudy skies of Europe they demanded that H_2S be fitted in their Fortresses and Liberators. Eventually as D-Day approached the possibility that H_2S systems would be of value to the Army emerged and the last chapters of this book deal with that problem, together with the hazardous issues raised when the Germans began using fast U-boats fitted with the Schnorkel breathing tube which enabled them to remain submerged when charging their batteries and with the anxieties about the use of the listening device Naxos used by the German night fighters to home on the bomber stream using H_2S.

Events of half a century ago are rarely remembered with precision; they are either forgotten or incorrectly remembered. When the European war ended I was in my 32nd year. I was released to return to the University in July 1945 and soon became so involved in the development of Jodrell Bank and astronomical problems that I quickly forgot the caldron of the war years and could not have written this book but for the fortunate circumstance that I had retained two documents of vital importance to the task. One was my personal manuscript diary. A good deal of this is concerned with entirely private and family matters but there is also the frequent reference to the state of the war and the occasional brief comments on my work. The quotations from this diary occur in the text as written—often with the odd phraseology of the war years and with sentences that might be better constructed in calmer days. Such quotations are inset with the date written and no additional references. Where the meaning may not be clear I have inserted an explanatory word in square brackets. The other document is a type-written account of H_2S, ASV and the associated developments compiled in the Telecommunications Research Establishment (TRE) after the capitulation of Germany in the two further months that elapsed before I could be released. These papers are clasped in a large hardboard file labelled Royal

Air Force Form 1740 and, fortunately, include several important copies of minutes of meetings that have proved of vital importance to the accuracy of the account in this book. The extracts from this document in the text are referenced to 'Lovell TRE record'. The record begins with the background to H_2S (Chapter 8) and continues to the end of the war. This document will in due course be deposited in the Imperial War Museum.

Apart from the early chapters about AI and the development of centimetre radar for that system the book is primarily concerned with the blind bombing aid, H_2S and the centimetre ASV derivation. Of the latter there can be no question of the crucial part it played in 1943 in the battle against the U-boats in the Bay of Biscay. The issues about the blind bombing aid H_2S are a different matter and are intricately involved with the whole question of the strategic bombing campaign of the RAF in the war years. A multitude of opinions have been expressed and many books have been written about this *per se*. I have made no attempt to add to this confusion of opinion but have concentrated on the task I was allotted, the technical solutions and the manner in which the device was used operationally. Those readers who wish to discover how this story of H_2S fitted into the overall strategic bombing campaign and the effects of that campaign should consult the four volumes of the official history *The Strategic Air Offensive against Germany 1939–1945*.[3] On the whole (and apart from a few minor inaccuracies of fact) this official account places fairly the use of H_2S within the broad strategic bombing campaign—and not always with praise. Several of the conclusions and statements in this official history have been severely criticised in *'Bomber' Harris* by Dudley Saward[4] who was the Chief Radar Officer in Bomber Command from 1941 until the end of the war. Readers of these two works will find authoritative but often diametrically opposed opinions about the effect of the strategic bombing offensive in World War II. My main purpose in the present book is to describe the technical background and the relevant political and operational attitudes to one of the scientific devices that was used by Bomber Command in this offensive.

References

1 Lovell Bernard *Astronomer by Chance* 1990 (New York: Basic Books); 1991 (London: Macmillan)
2 Bowen E G 1987 *Radar Days* (Bristol: Adam Hilger)
3 Webster Sir Charles and Frankland Noble 1961 *The Strategic Air Offensive against Germany 1939–1945* 4 vols (London: HMSO)
4 Saward Dudley 1984 *'Bomber' Harris* (London: Cassell)

Author's Note

I am indebted to several wartime colleagues for help, especially to those who have so willingly made suggestions about parts of the text describing the work in World War II with which they were so closely involved. Mr J R Atkinson, Professor W E Burcham, Sir Samuel Curran, Flight Lieutenant John Dickie, Professor R Hanbury Brown, Air Vice Marshal P M S Hedgeland, Sir Alan Hodgkin, Dr B J O'Kane, and Dr F J U Ritson have read various parts of the original text and the final form owes much to their advice. I have also received valuable help in correspondence with Mr R G P Batt, Sir Robert Clayton, Commander Derek Howse, Sir Andrew Huxley, Flight Lieutenant Len Killip, Dr Denis Robinson, and Group Captain Dudley Saward. I am indebted to the Director RSRE, Great Malvern, for permission to use the library and especially to the Senior Librarian RSRE, Miss Janet Dudley, and her staff for invaluable help in the search of the wartime archives. Whilst working there my contact with Mr W H Sleigh and Dr E H Putley, former members of TRE, led to further valuable sources of information. I am also indebted to those who have given permission to use illustrations or diagrams as indicated in the captions. Where no such reference is given the illustrations have been copied from 'Lovell TRE record' (see note in Preface) or are in the author's wartime collection.

Chronology of radar developments described in this book in relation to major wartime events

Date	Radar Development	Aspect of War
1935 January 28	First meeting of Tizard Committee	
February 26	Demonstration using Daventry transmitter	
June	First radar echoes at Orfordness	
1936 March	Move of Orfordness group to Bawdsey Manor	
1938 September	Thames estuary covered by radar chain (CH)	Munich crisis and agreement
1939 Spring	CH radar coverage of east and south coasts in operation	
September 1		Germany invaded Poland
September 3	Move of Bawdsey group to Dundee—became Air Ministry Research Establishment (AMRE). Bowen's airborne radar group established at Scone near Perth.	British declaration of war on Germany
Autumn	Fitting of $1\frac{1}{2}$ metre AI in Blenheim night fighters	
October 14		*Royal Oak* torpedoed by U-boats in Scapa Flow

(*continued*)

Date	Radar Development	Aspect of War
1939 November	Bowen's airborne group moved from Scone to St Athan in South Wales. Fitting of $1\frac{1}{2}$ metre ASV in Hudson aircraft	German battleships escape detection
December 13		Battle of the River Plate. Scuttling of *Admiral Graf Spee* (December 17)
1940 February 21	Randall & Boot 10 centimetre cavity magnetron first operated at University of Birmingham	
April 9		Germany invaded Denmark and Norway
May 7	AMRE moved from Dundee to Worth Matravers joined by airborne group from St Athan—became MAPRE (Ministry of Aircraft Production Research Establishment)	
May 10		Germany invaded Belgium and Holland Churchill replaced Chamberlain as Prime Minister
May 27–June 4		Evacuation of 338 000 of the BEF, French and other Allied troops from Dunkirk

Date	Radar Development	War Events
June 10		Italy declared war
June		Fall of France (Paris, June 14)
July 19	First Randall–Boot/GEC cavity magnetron arrived at Worth Matravers	Luftwaffe began daytime attacks on England (July 10)
August 12	First 10 centimetre radar echoes from aeroplane	
August 13	Echoes from tin sheet carried by Assistant on bicycle	Major phase of Battle of Britain (August–September)
September	AMRE moved from Worth Matravers to Leeson House (Langton Matravers)	Germans began night bombing of London (September 7)
September 13		Italy invaded Egypt
October 18	First AI night fighter operations under GCI control	
October 28		Italy invaded Greece
November 11	Demonstration to naval personnel of detection of 10 centimetre radar echoes from submarine in Swanage Bay	Night bombing attacks spread to cities other than London (November 3); Coventry raid (November 14)
November	MAPRE became TRE	
December 15		All Italian troops cleared from Egypt

(continued)

Date	Radar Development	Aspect of War
1941 January		Allies captured Tobruk (January 21/22) and achieved victory in the desert
March	First airborne tests of 10 centimetre AI using spiral scanner	Lease-lend bill signed by President Roosevelt Intensification of U-boat attacks on shipping in North Atlantic
April 6		Germany invaded Greece and Yugoslavia
April 24		Surrender of Greece and British withdrawal
May 27–30		British withdrawal from Crete
May 27		*Bismark* sunk
January–May	Increasing supremacy over German night bombers of $1\frac{1}{2}$ metre AI Mark IV in Beaufighters under GCI control	German night bomber losses rise to 10% of sorties. End of concentrated night attacks on English cities
June 22		Germany invaded Russia
September 3		Prime Minister asked Chief of Air Staff to give urgent attention to failure of RAF night bombing campaign

October 26	TRE meeting on night bombing problems	
November 1	Test of airborne 10 centimetre system in Blenheim for town detection	
November 14		*Ark Royal* sunk
December 7		Japanese attacked Pearl Harbour and USA entered war
December 10		*Prince of Wales* and *Repulse* sunk by Japanese
December 23	Secretary of State for Air ordered test flights to determine whether specific ground targets could be detected by airborne 10 centimetre radar	
1942 January 21		Rommel began counter attack in Western Desert
February 12		*Scharnhorst* and *Gneisenau* escaped from Brest to Kiel through English Channel
February 15		Fall of Singapore
February 27–28	Bruneval raid to capture German radar	
March 8–9	First operational use of Gee over the Ruhr	
March 27	Halifax V9977 equipped with cupola for H_2S scanner arrived at Hurn airport. Mark VII 10 centimetre AI operational in Beaufighters	

(continued)

Date	Radar Development	Aspect of War
1942 April 16	First flight of Halifax V9977 with H_2S	
May 25–26	TRE moved to Great Malvern	
May 30		1000 bomber raid on Cologne
June		Leigh Light equipped Wellingtons with $1\frac{1}{2}$ metre ASV began operating over Bay of Biscay
June 7	Crash of H_2S Halifax V9977 in South Wales	
June 21		Axis powers captured Tobruk with 33 000 Allied soldiers and began advance on Egypt
		Churchill/Roosevelt agreement to pool UK and USA knowledge and resources for development of atomic bomb
July 3	Prime Minister's meeting demanded two squadrons of H_2S by the autumn	
July 15	Secretary of State for Air ruled that H_2S should use the magnetron and that development of klystron version should cease	

August		U-boats fitted with $1\frac{1}{2}$ metre listening receiver (Metox) began to counteract effect of Leigh Light and $1\frac{1}{2}$ metre ASV over Bay of Biscay
September 30	First H$_2$S equipped Halifax to BDU for service trials	
October 23		Battle of Alamein began. Allied breakthrough and retreat of Rommel (November 4)
November 8		Allied invasion of French North Africa
November 20	First RAF squadron equipped with main production version of 10 centimetre AI (AI Mark VIII)	Siege of Malta raised
December		
December 20	First operational use of Oboe	
December 31	24 H$_2$S aircraft (12 Halifax and 12 Stirling) in 8 Group	
1943 January		Battle of Stalingrad
January 29		Surrender of von Paulus and German sixth army to Russians
January 30–31	First operational use of H$_2$S by Pathfinder Force in attack on Hamburg	

(continued)

Date	Radar Development	Aspect of War
1943 March 1	First operational use over the Bay of Biscay of Leigh Light Wellingtons equipped with 10 centimetre ASV (ASV Mark III)	
April–May		Dönitz orders U-boats to stay on surface and fight. 56 U-boats destroyed (30% of those at sea). Survivors withdrawn from North Atlantic
May	Mass production of H$_2$S (Mark II) began for equipping main bomber force	
May 12		Surrender of Axis powers in North Africa
July	Tank landing craft equipped with H$_2$S	Used in invasion of Sicily, July 10
July 23–August 3	Pathfinder Force with H$_2$S led massive raids on Hamburg	Hamburg destroyed (project Gomorrah)
September 3		Invasion of Italian mainland
September 7		Surrender of Italy
Autumn		Evidence of U-boat use of 10 centimetre listening receiver, Naxos

Date		
October	Fishpond (tail warning addition to H$_2$S) operational	
November 27	First operational use by Fortresses of US 8th Air Force equipped with H$_2$S Mark II	300 Fortresses attacked Emden
November–December	Six experimental 3 centimetre H$_2$S in Lancasters with 8 Group	3 centimetre H$_2$S in 8 Group led major raids on Berlin and Leipzig
1944	Introduction of improved versions of H$_2$S Mark II (10 centimetre) and Mark III (3 centimetre) to Pathfinder Force and Main Force	German night fighters equipped with Naxos listening device for homing on to H$_2$S transmissions
January 22		Allied landings at Anzio
June 4		Allies entered Rome
June 6		*Overlord*—Allied landings in Normandy—D-day
June 25	First flight of experimental K-band H$_2$S (Mark VI)	
July	Flight trials of engineered prototype of H$_2$S Mark IV	Schnorkel equipped U-boats first operational
July 4	First airborne test of 3 centimetre H$_2$S with 6 foot scanner (H$_2$S Mark IIIC: Whirligig)	
July 8	Demonstration to Bomber Command of 6 foot X-band H$_2$S systems (Mark III series) and of 3 foot X-band H$_2$S Mark IV	

(continued)

Date	Radar Development	Aspect of War
1944 August 15		Allied landings in Southern France
December	Demonstration of 6 foot K-band H_2S high definition reconnaissance system for Army collaboration	
1945 May 8		Surrender of German armies—VE Day
August 6		Atomic bomb dropped on Hiroshima
August 9		Atomic bomb dropped on Nagasaki
August 15		Surrender of Japan—VJ Day
Autumn	All versions of H_2S declared obsolescent except H_2S Mark IIIG (improved 6 foot X-band) and Mark IV/IVA	
1946	Flight tests of H_2S Mark IVA (6 foot X-band with computing facilities)	
1947–1950	H_2S Mark IVA standard fitting in Bomber Command Lincolns. Development of H_2S Mark IX/IXA for V-bombers	

1951	First flight trials of Mark IXA (6 foot X-band linked with navigation and bombing computer)
1956	Mark IXA in service in V-bombers	Valiant V-bombers using Mark IXA made pre-emptive strike against Egyptian Air Force at Cairo airport
1982 May 1	Last operational use of H_2S systems	Vulcan V-bomber using H_2S IXA and Navigation and Bombing system bombed Port Stanley airport in the Falklands

Chapter 1

August 1939

I do not know whether No 43 Albermarle Crescent in Scarborough still stands but in 1939 it was a boarding house of Victorian vintage. I arrived there on the Sunday evening of 27 August. That detail of the place in which I slept 50 years ago is not, in itself, of any importance to the story in this book. On the other hand the fact that I can provide these details with confidence is important. In those days of youth I wrote a diary. The daily entries are of no literary interest; but they are the factual—and often emotional—record of the years of war that would otherwise be tangled in the memory.

I did not stay in Scarborough for many nights but during my stay there the declaration of war was made. It was the expectation during the summer months of 1939, that this declaration of war was inevitable, that had led to my journey to Scarborough and I must first explain the background to this episode.

Staxton Wold

My brief visit to Scarborough and all that occupied me during the following six years of war, and indeed, for the remainder of my subsequent career, had its origins in 1934. At that time the mounting concern about the protection of the country against air attack led H E Wimperis, the Director of Scientific Research in the Air Ministry, to seek the advice of A V Hill the distinguished Cambridge physiologist, and a gunnery officer in World War I, about possible means of destroying enemy aircraft. The important result of these conversations was that on 12 November 1934 Wimperis sent a memorandum to Lord Londonderry, who was then Secretary of State for Air, advising him to set up a committee 'to consider how far recent advances in scientific and technical knowledge can be used to strengthen the present methods of defence against hostile aircraft'.

Wimperis suggested that the Chairman of the Aeronautical Research Committee, H T Tizard should be the chairman of the committee and that the other members should be A V Hill and P M S Blackett 'who was a Naval Officer before and during the War, and has since proved himself by his work

1

at Cambridge as one of the best of the younger scientific leaders of the day'. Lord Londonderry acted quickly on this advice from Wimperis.[1] Tizard received the formal invitation to be the chairman on 12 December and on the morning of 28 January 1935 Tizard, Hill, Blackett and Wimperis, met in the rooms of the Air Ministry with A P Rowe†, an assistant to Wimperis, acting as secretary. A great deal has been written about the affairs of this Tizard Committee[2], particularly in relation to the complex web of political circumstances that led to the appearance of F A Lindemann‡ (afterwards Lord Cherwell) as an additional member of the committee on 25 July and the subsequent resignation of Hill, Blackett and then Tizard. Blackett gave his own account of the crisis when he delivered the Tizard Memorial Lecture in 1960.[4]

> Although the Air Defence Committee started up in January 1935 with only Tizard, Hill, Wimperis and myself as members, in July of the same year the Secretary of State for Air, under pressure from Mr. Winston Churchill, enlarged it by the addition of Professor F A Lindemann (afterwards Lord Cherwell). It was not long before the meetings became long and controversial: the main points of dispute concerned the priorities for research and development which should be assigned to the various projects which were being fathered by the Committee. For example, Lindemann wanted higher priority for the detection of aircraft by infra-red radiation and for the dropping of parachute-carrying bombs in front of enemy night bombers, and lower priority for radar, than the other members thought proper. On one occasion Lindemann became so fierce with Tizard that the secretaries had to be sent out of the committee room so as to keep the squabble as private as possible. In August 1936, soon after this meeting, A V Hill and I decided that the Committee could not function satisfactorily under such conditions; so we resigned ...

After the outbreak of the war the increasing political influence of Churchill on his return to the Admiralty and then as Prime Minister, coupled with his use of Lindemann as his scientific adviser, led to the decease of this Tizard Committee. However, even before the first conflict with Lindemann in July 1935 the Committee had taken the vital steps that led to the development of radar (originally known as RDF—radio-direction-finding). One of the many ideas placed before the Tizard Committee at its first meetings was the vague concept of the 'death ray'. The Superintendent of the Radio Research Station§ at Slough—R A Watson-Watt—was asked for

† A P Rowe had been appointed an Assistant II to Wimperis in 1922.
‡ The inclusion of Lindemann as a member of the original committee was discussed and rejected.[3]
§ For the origin of the Radio Research Station under the Department of Scientific and Industrial Research (DSIR) in 1927, its relation to the National Physical Laboratory and the circumstances leading to Watson-Watt's appointment as Superintendent see J A Ratcliffe.[5]

an assessment. He asked A F Wilkins, a member of his staff, to calculate how much radio energy would be needed to damage an aircraft or its crew. There was no such possibility but in making the calculations Wilkins concluded that if an aircraft flew through a beam of radio waves then it would scatter back to earth sufficient energy for it to be detected. The appropriate memorandum† from Watson-Watt reached the Tizard Committee in mid-February and the subsequent demonstration using the Daventry transmitter on 26 February 1935 immediately triggered the sequence of events that led to the defensive chain of radar stations along the east and south coasts of England.

E G Bowen, one of the four pioneers who began the erection of the first experimental radar on a remote site at Orfordness, 90 miles north-east of London, has described the events that followed the February demonstration using the Daventry transmitter.[8] In mid-June 1935 only one month after arriving at Orfordness they obtained radar echoes from a flying boat at a range of 17 miles, by September the range on aircraft had been increased to 40 miles, and by the end of 1935 they had increased the range of detection to 80 miles.

The success of these developments at Orfordness led the Air Staff to ask for a group of five radar stations to cover the approaches to the Thames estuary. The Treasury allocated £10 000 000 for this work and by the time of the 1938 Munich crisis this early warning network was in operation. About the influence of the Tizard Committee Blackett wrote:[9]

> The full backing of the Committee became an effective way of getting high priority given to a project. Without such mutual trust the scientific development of radar could not have forged ahead at such speed, nor would the tens of millions of pounds have been provided so rapidly and secretly by the Treasury to build the radar chain.

By Easter 1939 the radar chain had been extended to cover the east and south coasts from Ventnor to the Firth of Tay and a 24-hour watch closely integrated with the headquarters of Fighter Command had commenced. Staxton Wold, the high ground a few miles inland from Scarborough was the site of one of these operational CH (Chain Home) radar stations.

At that time I was an assistant lecturer on the staff of Blackett's physics department in the University of Manchester and it may well be asked what business I had to be in this highly secret radar station at that moment. The

---

† The first memorandum in which it was concluded that there was no possibility of damaging the aircraft or crew was circulated to the Tizard Committee on 4 February 1935. On 6 February A P Rowe (as Secretary of the Committee) asked Watson-Watt for his further assessment of the possibility of detecting the aircraft by radio means.[6]

The second memorandum *Detection and location of aircraft by radio methods* was submitted by Watson-Watt on 12 February. It is reprinted in his book *Three Steps to Victory*.[7]

answer lies in a typically British method of making important arrangements. Early in 1938 Tizard asked J D Cockcroft of the Cavendish Laboratory to join him for luncheon in *The Athenaeum*. Tizard wanted his help and having explained about the secret development of radar said that the device would be troublesome to maintain and asked Cockcroft if, in the event of war breaking out, some of the physicists from the Cavendish would help. However, Tizard could not obtain immediate authority to release the radar secrets to groups of University scientists and on 13 September 1938 he wrote to W L Bragg, the Cavendish Professor of Physics, that a decision about the introduction of scientists to the radar secrets was put off until the autumn but asked for lists of people who might be prepared to commit themselves.[10]

As the tensions in Europe developed in 1939 these reservations disappeared and in February of that year Watson-Watt visited Cambridge and, at a meeting chaired by W L Bragg and attended by four senior members of the Cavendish staff, a list of scientists was drawn up and divided into groups to be allocated to the various CH stations. I knew nothing whatsoever of these arrangements or, indeed, of the existence of radar until Blackett telephoned from London on 29 July urging me to abandon my plans for driving a van-load of cosmic ray equipment across France to the Pic du Midi in the Pyrenees.

A few weeks earlier, on 10 July, I had arrived in the physics department to find Blackett 'looking like death' waiting to see me. I was to be sent to help at a secret Air Ministry station in September. He gave no explanation or any indication of the work involved or of the function of this Air Ministry station. There was a little more elucidation in his call on 29 July—the secret station was somewhere on the Yorkshire coast, I was to be 'technical chief' of a group from Manchester and I was to report to Bawdsey Manor on the east coast on 14 August. Life became a strange mixture of normality and apprehension. In the physics department I was in the thick of some fascinating research, using an expansion chamber to photograph cosmic ray showers of high energy. During the day I was torn between that and the cricket at Old Trafford where the test match between England and the West Indies was in progress. Nearly two years earlier I had married Joyce Chesterman and in the evenings we would play tennis or drive over the hills to a theatre in Buxton or to a cinema in Manchester, seeking relief from the fear of the future but blissfully unaware of the rigour of the years that lay ahead. But that life of a precarious peace could not last. On 4 August I sadly closed down the expansion chamber, little realising that it would be many years before I would return to that kind of research. The dreams and life of peace and youth vanished. On that Sunday evening of 13 August I arrived in Felixstowe, separated by a river's breadth from Bawdsey Manor.

The small group at Orfordness under E G Bowen and A F Wilkins had moved to Bawdsey Manor on the Suffolk coast in March 1936. Watson-Watt was appointed Superintendent of this Bawdsey Research Station. Initially run by the National Physical Laboratory (at that time a laboratory of the Department of Scientific and Industrial Research (DSIR)) it was transferred to the Air Ministry on 1 August 1936 and with the development and construction of the coastal radar CH stations this Bawdsey research group grew rapidly—and so did the Air Ministry organisation. In May 1938 Watson-Watt was appointed Director of Communications Development (DCD) at the Air Ministry, and A P Rowe succeeded him as the Superintendent at Bawdsey and it was he who greeted a dozen of us from various universities on the morning of 14 August 1939.†

This was the first time I had seen or even heard of Rowe. He was of small physical stature, wore horn-rimmed spectacles and smoked a pipe. I noticed a cricket bag in the room in which he greeted us but I was soon to learn that this was no symbol of the way in which Rowe ruled that establishment for the next six years. On that morning our encounter was brief. Soon he handed us over to W B Lewis, whom Cockcroft, Ratcliffe and the others from Cambridge clearly knew well. Nearly 50 years later, when I had the task of writing part of the Lewis biographical memoir[11] for the Royal Society, I realised that Lewis was then a new recruit to Rowe's staff at Bawdsey. For some years previously Lewis had worked with Rutherford and with Cockcroft at the Cavendish Laboratory and had particular skills in radio and electronics. In July 1939 he had been seconded to the Air Ministry by the Cavendish Laboratory—nominally for a period of two years—but he was never to return. He was to be a cardinal influence in the task of assimilating people like myself from the universities who were soon to join Rowe's staff in considerable numbers.

The three days we spent at Bawdsey Manor under Lewis' guidance revealed to me an entirely different order of science and technology compared with that of my university experience. Even more electrifying were the visits that followed to the headquarters of Fighter Command at Bentley Priory, Stanmore, and then to the RAF station at Biggin Hill. In the underground filter room at Stanmore we had seen the reaction to a mock raid by a French squadron—the radar plots from the CH stations guiding the controllers to direct the appropriate fighter squadrons, and then at Biggin Hill the scrambling of a Hurricane squadron and a mock interception of enemy bombers. The great achievement that was to become manifest in the 1940 Battle of Britain had been revealed to us in those few days. It was not merely the radar detection of approaching bombers but the integration and efficient use of that information with the operational command that was to prove of such vital importance. With any remaining hope that war would not come

---

† Amongst others in this group where J D Cockcroft, J A Ratcliffe, M V Wilkes, N Feather, J T Randall, C E Wyn Williams, H W B Skinner, G E F Fertel and L G H Huxley. All strangers to me except Skinner who was on the staff at Bristol where I had been a student.

shattered by the news of the German–Soviet pact, I reported to Staxton Wold on the morning after my arrival in Scarborough. Our arrival there coincided with that of more than a hundred army personnel whose task was to mount anti-aircraft guns and defend the radar station against attack. We were very much in the way and no one seemed to know how we were supposed to be helping this efficiently-run establishment. Eventually the RAF Warrant Officer in charge agreed that we could study the technical and operating instruction manuals and take shifts with the various operators. The most obvious feature of the CH station on Staxton Wold was the magnificent 360 foot steel towers carrying the transmitter aerials. There were six half-wave dipoles, stacked vertically with a tuned reflector curtain. This aerial system produced a polar diagram with a horizontal beam-width of about 60 degrees effectively 'floodlighting' the area in front of the station. The receiving aerials were on separate 240 foot wooden towers and consisted of two pairs of crossed dipoles at different heights. The dipoles in each pair were mounted at right angles to each other, one of them being perpendicular to the axis or 'line of shoot' of the transmitted beam. By comparing the strength of the radar echo in the two aerials by the use of a goniometer coil it was possible to estimate the bearing of the aircraft. Similarly a comparison of the echo strength from the dipoles at the different heights gave an approximate indication of the height of the aircraft.

At the time of my visit to Staxton Wold† there were three sets of these crossed dipoles on the wooden receiver towers at heights of 215 feet, 95 feet and 45 feet above the ground. If the target aircraft was at low elevation comparison of the echo amplitudes was made between the dipoles at 215 feet and 95 feet, whereas, for aircraft at high elevation, the dipoles at 95 feet and 45 feet were used. The former system produced a polar diagram with the axis about 14 degrees to the horizontal while the 95 foot and 45 foot system had a maximum lobe at about 48 degrees. In the systems used in 1939 the same goniometer was used for determining bearing and height. When the operator desired to estimate the height he pressed the appropriate button on the control desk and this resulted in the top dipoles feeding one field coil of the goniometer used for direction finding whilst the lower dipole system fed the other field coil. Calibration was made by using balloons, autogyros or small aircraft and the normal practice was to turn the position of the goniometer until the echo strength was a minimum (the echo signal was 'D/F'd out'). The pointer reading then gave the bearing, or height, from the previous calibrations. The bearing and height errors from any single CH station using this system were considerable (bearing errors as much as 12 degrees were

† The CH station at Staxton Wold became operational on 1 April 1939. A detailed history of the construction and wartime operation of the station has been published by Flt Lt Peter Emmett[12] (a post-war Station Commander at Staxton Wold). I am indebted to Flt Lt Emmett for drawing my attention to this publication.

common) but the combination of the data in the Stanmore filter room from pairs of adjoining stations led to a most efficient system. One major problem was that CH stations such as the one at Staxton Wold would not detect aircraft at angles of elevation less than about 1½ degrees and a home defence exercise in August 1938 had confirmed that enemy aircraft could avoid radar detection at any operationally useful range by flying towards the coast at low level. This problem had been explained to us at Bawdsey when we were shown the prototype of a supplementary radar known as CHL (Chain Home Low) to be used at the main CH stations. Initially the CHL installations consisted of two sets of broadside arrays rotating in synchronism (one transmitting and the other receiving). These worked on a wavelength of 1½ metres and produced a narrow beam at low elevation. At the time of my visit to Staxton Wold no CHL stations were operational but the failure to detect enemy mine-laying aircraft led to their urgent production. Eleven were operational by mid-February 1940 and a further seven by the end of that month. The original prototype CHL station was to undergo rapid development and various modifications soon came into use for coastal defence, and, as we shall see later, for ground controlled interception (GCI).

Although the aerial towers were the most obvious part of Staxton Wold the powerful radar transmitters and the receiving and display equipment were examples of the most advanced techniques of that era, far beyond anything of my previous experience. The first system at Orfordness in 1935 worked on a wavelength of 50 metres in the belief that the wing-span of a bomber would act as a dipole resonator on this wavelength and would give maximum signal strength.† There was so much interference from commercial radio traffic on this wavelength that Bowen and his colleagues were soon forced to reduce the wavelength of operation, first to 26 metres and then to the 10 to 13 metre band. In March 1938 the Bawdsey research group drew up the specification for the transmitters to be installed on the main CH stations. These were to operate in this wavelength region (more specifically in the frequency band 20–55 megahertz) and were to be capable of transmitting pulses of radio waves with a power exceeding 200 kilowatts. The pulse repetition rate was to be either 25 or 50 per second with a pulse width capable of adjustment within the range 5 to 35 microseconds.

The development and manufacture of the transmitters to meet these specifications was undertaken by Metropolitan Vickers, and J M Dodds[13] has given a detailed description of the problems encountered and their solution. The transmitter valves were demountable, continuously evacuated and water-cooled. The receiving equipment and operating desks, including the cathode-ray tube display designed and built by A C Cossor, have been described by J W Jenkins.[14] A special commutator-type range switch was mounted on front of the cathode-ray tube and a similar switch was coupled

---

† The wing-span of the bomber of that period was about 75 feet.

to the goniometer shaft. These, feeding range, bearing and height into an analogue computer device enabled the operator to transmit the grid reference of an approaching aircraft direct to the filter room at Stanmore.

This massive equipment and advanced technology had a dramatic effect on me as a young man emerging from one of the most advanced university laboratories of the day where Bunsen burners were still in use to heat soldering irons. On 1 September, when the news came that Germany had invaded Poland, I was getting direction-finding practice on aircraft flying inland and, on the Sunday morning of 3 September as our ultimatum to Hitler expired, I was watching the WAAF operator at the control desk. The cathode-ray tube (CRT) was full of echoes and I was surprised that she did not perform the routine tasks with the range switch and goniometer to transmit the positions to Stanmore. She explained that the echoes were not from aircraft because they were only transient and that they referred to them as ionospheric. I have explained elsewhere[15] how the sight of those echoes eventually led to the emergence of the radio telescope and the observatory at Jodrell Bank.

The next day I was with the night shift. We were at war but there was not a sign of an aircraft echo on the tube. After the expectation of immediate bombing raids, this was quite uncanny. The RAF was dropping millions of pamphlets over Germany but the reality of war was manifest by the news that the *Athenia* had been torpedoed and that a thousand people had been rescued from the sinking ship. Life was pleasant enough in almost empty, blacked-out Scarborough, but I was longing for a real job to do and when Blackett came to see us on the seventh he promised to get me into one of the radar research groups. I left Staxton Wold and Scarborough on 9 September.

References

1 The note of the discussion which Wimperis had with A V Hill, submitted to the Secretary of State, on 12 November 1934, is in the PRO (Public Record Office) file AIR2/4481 which also contains the subsequent exchange of minutes leading to the despatch of the letters of invitation to Tizard, Hill and Blackett on 11 December 1934

2 See, for example, Clark R W 1965 *Tizard* (London: Methuen), Snow C P 1961 *Science and Government* (Oxford: Oxford University Press), and 1962 *A Postscript to Science and Government* (Oxford: Oxford University Press)

3 Minutes of meeting in PRO file AIR2/4481

4 Blackett P M S 1960 *Tizard and the Science of War* (Institute of Strategic Studies, 11 February 1960); 1960 *Nature*: 185, 647; reprinted in Blackett P M S 1962 *Studies of War* pt 2 Chapter 8 p 101 (Edinburgh: Oliver & Boyd)

5 Ratcliffe J A 1975 *Robert Alexander Watson-Watt* in *Biog. Mem. R. Soc.* 21 549

6 Air Ministry File Misc. Papers S35227 in PRO file AIR2/4482

7   Watson-Watt R A 1957 *Three Steps to Victory* (London: Odhams) appendix p 470

8   Bowen E G 1987 *Radar Days* (Bristol: Adam Hilger)

9   See reference 4

10 Tizard Papers HTT125 Imperial War Museum

11 Lovell Bernard and Hurst D G 1988 *Wilfrid Bennett Lewis* in *Biog. Mem. R. Soc.* **34** 453

12 Emmett Flt Lt Peter July 1989 *Echoes of Victory* in *Beacon News*—the magazine of the RAF at Staxton Wold p 8 *et seq*

13 Dodds J M and Ludlow J H 1946 *The CH radio location transmitters* in *J. Inst. Elect. Eng.* **93** pt IIIA 1007

14 Jenkins J W 1946 *The development of CH type receivers for fixed and mobile working* in *J. Inst. Elect. Eng.* **93** Pt IIIA 1123

15 Lovell Bernard 1968 *The Story of Jodrell Bank* (Oxford: Oxford University Press and New York: Harper & Row); 1990 *Astronomer by Chance* (New York: Basic Books); 1991 (London: Macmillan)

Chapter 2

Scone Airport

In those first September days of the war the official Air Ministry appointing procedures suffered some disarray. Those of us who had been introduced to Bawdsey and had spent some time on CH stations expected to be summoned immediately to urgent wartime tasks. But this did not happen and like others of that group I spent the first two or three weeks of the war without an official appointment or instructions. In fact, it seemed likely that I would soon be unemployed since my university appointment expired at the end of September and by abandoning the proposed research on the Pic du Midi I had forfeited the intended support from the Department of Scientific and Industrial Research (DSIR). It was an aggravating situation and I returned to Manchester, blacked out, and packed up our first married home and pestered Blackett for action. At last on 27 September—four days after petrol was rationed—I was instructed to report to Dundee as quickly as possible.

The Bawdsey research group had moved from Bawdsey Manor to Dundee in the first few days of September. Rowe was apparently convinced that Bawdsey would be a prime target for the German bombers and Watson-Watt with his Scottish connections had arranged with the Vice-Chancellor of Dundee University to provide emergency accommodation. The arrival of the Bawdsey team—now officially the Air Ministry Research Establishment (AMRE) has been described by Bowen. [1]

> When he arrived at Dundee, Rowe was greeted by an incredulous Vice-Chancellor who seemed to have forgotten Watson-Watt's visit; he had simply not been informed of any subsequent plans. Being an accommodating soul, he offered the new arrivals two rooms, each 20 feet square, which was all the space he could spare for the several hundred people who were either en route to Dundee or already there. I do not want to elaborate on how the problem was finally solved and will leave those responsible to explain what transpired as best they can. There had been a monumental blunder, from which the Air Ministry Research Establishment, as it was now called, took a year or more to recover.

I had been instructed to report to W B Lewis in Dundee but when I arrived there on 29 September he was away and it was Sidney Jefferson, one of the senior members of the original Bawdsey group, who saw me and explained that I was not to be allocated to the research establishment in Dundee but to Taffy Bowen's airborne research group at Scone airport, near Perth.

More than 40 years later I was concerned with writing the history of the Anglo–Australian Telescope built at Siding Spring in New South Wales in the 1960s.[2] A critical phase of the negotiations about this joint enterprise occurred during the time when Blackett was President of the Royal Society and in searching the relevant Blackett archives in the Royal Society, I discovered, entirely by accident, the Blackett correspondence which led to the arrangement for me to join Bowen's group. After his visit to us at Staxton Wold he had attempted and failed to make contact with Cockcroft about arrangements for my wartime research. Then on 19 September he wrote directly to Lewis suggesting that I should go to Bowen for work on AI (Air Interception): 'I would be rather pleased as it then would give him a personal contact with the branch of RDF in which I am particularly interested in connection with the possibility of blind bombing from above'. Lewis replied on the twentieth. 'Come at once—welcomed by Bowen at Scone airport near Perth'.[3]

Air Interception (AI)

As it happened, Blackett's reference to blind bombing† was a prediction of my major wartime task, but that lay years ahead and when I arrived at Scone airport Bowen and his group were entirely occupied with the problems of AI. It is a remarkable tribute to the insight of Tizard and his committee that the need for an airborne radar interception device had been foreseen several years before the outbreak of war. Early in 1936 the experimental radar at Orfordness was achieving ranges of over 100 miles on aircraft. Tizard and his colleagues were satisfied that, given the development and integration of the system with Fighter Command, the daytime attacks of the German bombers could be defeated, and that the German Air Force would then switch to night bombing. That it did so in the winter of 1940–41 is a matter of history. In 1936, having foreseen this development, the Tizard Committee recognised that this would present a formidable problem for the RAF. Under good conditions at night, a maximum range at which a fighter pilot could make a visual identification of his target was 1000 feet or less, and the ground-based CH radar chain could not possibly provide positional information of that precision. The Committee pressed Watson-Watt, who was by

† I believe that Blackett had in mind the defence against German bombers by attacking them from above by blind bombing and not the blind bombing of ground targets.

then the Superintendent of the Bawdsey Research Station, to begin the development of a miniaturised radar which could be carried in a night fighter and home onto the suspected target from a range of about four miles.

When the Orfordness group moved to Bawdsey in the spring of 1936 the task of devising an airborne radar for this purpose was given to E G Bowen. He has described the pioneering work of his small group on this formidable problem.[4] When we were introduced to the secret of radar in August at Bawdsey there seemed to be some question as to whether we would be informed about this development. Eventually Rowe impressed on us that we were about to see a 'secret within a secret' and that was the first occasion on which I met Bowen with whom I was soon to suffer many tribulations.

At that time Bowen had an experimental airborne radar working on a wavelength of 1½ metres which he had demonstrated to several people, including the C-in-C Fighter Command. This equipment was fitted either in an Anson or Battle aircraft and, no doubt, under normal circumstances there would have been time for further development to overcome many problems and to make the radar installation in an aircraft more suitable for night inter-ception of enemy aircraft. The Air Ministry decided to adopt the twin-engined Blenheim aircraft as a night fighter and by the end of July 1939 Bowen had surmounted some of the problems in the conversion of the air-borne radar from the single-engined Battle to the Blenheim. Hitherto, dis-cussions had centered on the possibility of equipping half a dozen Blenheims for further trials to develop a prototype AI radar that could ultimately form the basis of a properly engineered version. Then at the beginning of August an emergency decision was made that 30 Blenheim aircraft must be equipped with the AI radar by 1 September. Orders had been placed for the transmit-ters and receivers with the firms who had been responsible for the main CH stations, namely Metropolitan Vickers and A C Cossor respectively, and Bowen has described in some detail the confusion and chaos that ensued. The decision to equip 30 night fighter Blenheim aircraft for operational use by 1 September had been made a month after a demonstration to the C-in-C Fighter Command. As Bowen wrote: 'the decision making had been rapid, but it bore little relation to the resources available to carry it out'.[5]

The entire problem had been exacerbated by the emergency move of the Bawdsey group to Dundee in the first days of September. So far, Bowen's group had used the RAF aerodrome at Martlesham Heath, well equipped and conveniently close to Bawdsey. When Bowen asked Watson-Watt about flying facilities at Dundee he was informed that arrangements had been made to use the airfield near Perth, at Scone. The manager of this small aerodrome was a civilian under contract to Training Command of the RAF to provide initial flight training in light aircraft. The facilities were, therefore, meagre, but by the time I arrived there at the end of September Bowen had per-suaded the manager to let him use one of the two small aircraft hangars in which he had arranged a very small space for laboratory work. Most of the fitting of the radar in the Blenheims had to be carried out in the open and I learnt straightaway that a move to a larger and better equipped aerodrome was to be expected soon.

The AI system being fitted into these Blenheim night fighters worked on a wavelength of 1½ metres. The small transmitter produced a peak power of about 2 kilowatts and was pulsed at 2000 per second, the pulse width being nominally about one microsecond. The transmitter was fed into a single dipole with reflector on the nose of the aircraft and this produced a broad beam 'floodlighting' the region ahead of the aircraft. Two other pairs of dipoles were mounted either side of the fuselage, or on the wings, and produced overlapping polar diagrams in azimuth and elevation. The echo from the target aircraft was displayed on two cathode-ray tubes, one for azimuth and the other for elevation. A four-position radio frequency switch, continuously changed the input of the receiver to the dipoles and a synchronised output switch displayed the strength of the echo from the pair of aerials side by side on the two cathode-ray tubes. When the target aircraft was detected the radar operator would instruct the pilot to change height and direction until the echo pairs were of equal amplitude. Under these conditions the fighter would be heading directly toward the target. The range of the target could be passed by the operator to the pilot together with this port/starboard, up/down information, in a continuous flow.

In the early experimental trials Bowen used a Battle aircraft, a single-engined fighter with a smooth fuselage. The azimuth and elevation aerials were mounted either side of the fuselage and suitable overlapping polar diagrams were obtained without great difficulty. The two-engined Blenheim presented a far more difficult problem for the azimuth aerials and this issue was still troublesome when I became acquainted with the system. The transmitter was extremely crude—two exposed thermionic valves mounted on a piece of wood and connected as a squegging oscillator. Bowen's group had fitted this AI Mark I equipment into six Blenheim night fighters and had despatched them to 25 Squadron at Northolt in August 1939. I arrived at Scone airport when the remainder of the 30 aircraft were being installed.

Apart from the poorly engineered version of this first AI there were fundamental problems with the system—one connected with the maximum range and the other with the minimum range. The maximum range problem arose because of the broad beam of the 1½ metre system. This floodlit the area in front of the fighter and radar echoes from the ground (ground returns) occurred at a range equal to the height at which the fighter was flying. At that time this was not more than about 15 000 feet. This placed a severe restriction on the range at which the radar in the night fighters could locate a target aircraft since the ground returns obscured the echo from the target aircraft at ranges greater than the height of operation above ground.

The main CH coastal network could position the invading aircraft with an accuracy which was adequate for the daytime fighter aircraft of the RAF to be directed to visual contact, but at night this was not the case. During the course of the pre-war tests of the AI system R Hanbury Brown, who had joined Bowen's airborne group, foresaw this difficulty. He pointed out that if the AI radars were to be of use in darkness, it was essential to develop a supplementary ground radar capable of a positional accuracy better than 15 000 feet, and preferably a system that gave instantaneous indication of the relative position of the fighter and the target aircraft. Bowen, backed by Fighter Command, made the appropriate request to the main establishment at Bawdsey. The advice of Bowen and Hanbury Brown was accepted and urgent attention was given to the development of a supplementary ground radar.

Initially the development was based on a system that had already been developed at Bawdsey to cover the deficiency of the poor range performance of the CH radar network on low-flying aircraft. This was the CHL (Chain Home Low) system working on a wavelength of 1½ metres. A considerable effort of the main research group at Dundee was involved in adapting the basic CHL system to make it useful for ground controlled interception (GCI), so that the night fighters could be directed sufficiently close to the enemy bomber to overcome the range limitations introduced by the ground returns of the AI radar. In adapting the CHL system two major features were introduced. First, the CHL radar gave azimuth position of the target but not height and a system had to be introduced to give this vital information. Second, although the beam of the CHL radars could be swung over a large sector, the presentation to the operator was fundamentally that of a cathode-ray tube range display with the azimuth of the aerial beam separately indicated. For the efficient direction of the night fighter it was essential to develop a display which gave instant information of the differing bearings of the fighter and the target. It was this requirement that led to the idea and development of the plan position indicator (PPI)—known originally as the radial time base because the time base of the cathode-ray tube rotated in synchronism with the aerial beam.

The first night fighter operations controlled by this GCI equipment took place during the night of 18 October 1940. Experience and training soon enabled the RAF to inflict considerable losses on the German night bombers. The first combat occurred during the night of 19–20 November 1940 and, by the spring of 1941, losses of 10 per cent were often inflicted on the German night bombers and this was the major factor leading to the cessation of this phase of the night bombing in mid-May 1941. † There were, however, major factors other than the GCI development that led to the operational successes with the AI equipment. The Blenheim aircraft could not compete

† Until February 1941, the German night bomber losses were less than 1 per cent of sorties and the anti-aircraft guns had the monopoly of success. From March 1941, the GCI/AI combination had the monopoly of success as the German losses increased to a level they could not sustain.

with the performance of the German night bombers and, by this time, the AI was installed in the more powerful Beaufighter aircraft which had replaced the Blenheims in the RAF night fighter squadrons.

Another major factor was the improvement of the AI system. The AI Mark I system, which I had first encountered when I arrived at Scone, had by this time evolved rapidly through AI Mark II, and Mark III to the AI Mark IV fitted in the Beaufighters. These successive improvements had revolved around the second fundamental problem of the AI Mark I—namely the minimum range performance.

The issue of the minimum range bedevilled the early development of the AI radar and was at the core of an increasing acrimony between Rowe and Lewis, whom Rowe had appointed as his deputy, and Bowen, which soon led to Bowen's eviction from the mainstream radar development and his departure to the USA in the autumn of 1940. The AI Mark I which was being fitted to the Blenheim aircraft when I joined Bowen's group in September 1939, had a poor pulse shape arising from the fact that it used a squegging transmitter and no modulator. This, and the breakthrough into the receiver which led to a temporary paralysis, limited the minimum range to about 1000 feet. This system had been demonstrated by Bowen to the C-in-C Fighter Command during the summer of 1939 who had expressed satisfaction with its performance. Further, this figure of 1000 feet at which a pilot could make identification of a target aircraft at night, had been established by night-time trials by the Royal Aircraft Establishment. There appears to be two reasons why the belief arose in AMRE in Dundee that this figure of 1000 feet was inadequate and that it was essential to reduce the minimum range to at least 600 feet. The first was representations from H Larnder, a former member of the Bawdsey staff and now with the operational research section of Fighter Command, that 1000 feet was quite inadequate for visual identification at night and the second was the general dissatisfaction with the poor performance and frequent breakdown of the first AI radars.

In any event, Rowe and Lewis in Dundee began to take urgent action to find some means of reducing this minimum range. Lewis introduced a device to damp out the tail of the transmitted pulse and, in collaboration with E H Cooke-Yarborough, first tested this system in the early spring of 1940. Eventually they succeeded in reducing the minimum range to 500 feet. Simultaneously, a contract had been placed with EMI where A D Blumlein and E L C White introduced a pulse modulator. This was also successful in reducing the minimum range to about 500 feet. The EMI system was eventually introduced as AI Mark IV since, in addition to reducing the minimum range, it gave an overall improved pulse shape and a superior maximum range. This urgent work was carried through without the collaboration and agreement of Bowen who, as will be seen, had become isolated with his group in South Wales during that winter of 1939–40.

The irony of this sequence of events is that the minimum range issue probably had little to do with the early operational failures of AI that led Tizard to press the need for detailed scientific trials. These were carried out in April 1940 at Tangmere by a special unit—the Night Interception Unit—later the Fighter Interception Unit (FIU). These trials carried out by Hanbury Brown[6] flying with Group Captain Chamberlain revealed that the major trouble was a squint of the AI aerial beams. It seems that this trouble arose because the original test installations were made on long-nose Blenheims whereas the operational squadrons used short-nose Blenheims. In any event when the squint difficulty was rectified the problem of the minimum range vanished.

A retrospective judgement on the evolution of the early versions of the AI radar is that the introduction of the modulator solved a problem that did not have operational importance. Before the war began Bowen and his group had solved the problems of receiver paralysis and they had discarded the modulator solution as unnecessary because of the evidence from the Royal Aircraft Establishment (RAE) that the minimum range of 1000 feet was adequate. In fact, subsequently Bowen made an assessment of the minimum ranges recorded by Fighter Command pilots in the 1940/41 combat reports. The median range to which enemy aircraft were tracked by the 1½ metre AI and then seen visually was between 1200 and 1500 feet.

The Winter of Discontent

Throughout the winter of 1939–40 I was little more than an onlooker on the fringe of these disputes. Initially my ardent desire to plunge into some useful work seemed likely to be satisfied. Another of my young colleagues, Peter Ingleby, from Manchester had arrived and Bowen said we could start work on the transmitters. He wanted us to push the wavelength down from 1½ metres to 50 centimetres. I doubt if I understood the full impact of this at that stage, but it was the beginning of the long haul to make an AI work on a short enough wavelength so that a narrow beam could be transmitted from the aircraft, thereby avoiding the ground returns which limited the range of detection to the height at which the aircraft was operating. A member of Bowen's group had shown me the installations in the Blenheim and I was full of excitement and hope.

That state of euphoria lasted only for 24 hours. The next morning, 2 October, when I arrived at the airport I was told there had been a change of plan and that I was to help to make the existing 1½ metre transmitter work correctly. Thereby began the see-saw existence—sometimes helping with the 1½ metre installations and sometimes attempting to produce a reasonable transmitter power on shorter wavelengths. Bowen and his deputy, Gerald Touch, exemplified two opposing schools of thought. Touch said we must realise there was 'a war on' and that all effort should go into getting the existing 1½ metre AI into operation. Bowen looked more to the future and to the possibilities for AI if the ground returns could be avoided.

Working in a small group with this type of conflict seven days a week (all leave and rest days had been cancelled) soon eroded the enthusiasm and hope. Furthermore, I had been allowed to join in a test flight of the AI and that quickly made me realise that making an equipment work in the air was quite a different matter from operating on the ground. Hitherto in my few years of university research I had donned a lab coat and dealt with an apparatus in the warmth and quiet of a laboratory. Now the lab coat was replaced by a bulky flying suit, a parachute harness and a parachute which also served as a seat. The spacious laboratory was replaced by the cramped and cold interior of a Blenheim night fighter. The noise made normal conversation impossible and the vibration was so great that I could not imagine how any electronic equipment could survive even on its anti-vibration mountings. I was soon to become accustomed to these hazards of airborne radar. At least on that first flight we levelled off at 10 000 feet and there was no oxygen mask to hinder the splendid view of Perth, the River Tay and the snow capped mountains glistening in the sunshine of that Sunday morning.

By mid-October I had progressed with Ingleby to the extent of making a transmitter work on a wavelength of 1 metre but the efficiency was poor and we wanted to consult the relevant literature in the library at Dundee—25 miles away. Touch said we could not go, it would be 'a waste of time' and that we should apply ourselves to the existing apparatus. At this stage, only two weeks after joining Bowen's group, I was so appalled at the general confusion, the constantly changing decisions, and the lack of a library and workshop facilities, that I wrote a long letter of complaint to Blackett†. The date was 14 October, a day darkened by the sinking of the *Royal Oak*, with 1200 men on board, by U-boats penetrating the defences of Scapa Flow.‡

A week later Blackett replied—a calming letter telling me to be more tolerant until I had accomplished something in these technical fields and to remember that 'all defence work is like all Service work, very much a matter of dealing with people—the qualities of the personnel are part of the experimental facts and it is no use getting too upset about them'. It was good advice and as far as I knew at the time that was the end of the matter. That this was not the case I discovered by accident more than 40 years after the end of the war. In 1987 W B Lewis died and I was asked by the Royal Society to write the part of his biographical memoir until he left for Canada in 1946.[8] For certain sections of this I wanted to discover the extent of his liaison with Tizard and was given permission to search the relevant section of the Tizard archives in the Imperial War Museum. Then I discovered that Blackett had been so concerned about the contents of my letter to him of

† The substance of this letter has been published by D Saward.[7]
‡ 899 officers and men were lost.

14 October 1939 that he had sent it to Tizard on 19 October, saying that it was from:

> ... a young assistant lecturer from Manchester who has recently gone to Perth
> ... the impression he has got of conditions at Perth is depressing. It is only
> the huge importance of AI which makes me pass on these grumbles from
> Lovell .... I thought that if you do pay a visit there the letter might be of use
> to you in suggesting things to look out for ... I would rather you destroyed
> Lovell's letter in case it got by mistake circulated to others ...

Tizard did not destroy the letter but without revealing the source either sent the essence of it to Rowe or discussed the contents with him during a visit to Dundee.

It is as well I knew nothing of this when Rowe summoned me to visit him in Dundee on 26 October. Rowe listened attentively to my many complaints. He had on one wall of his office a large magnetic board with the names of all his staff on magnetic strips which could readily be moved from place to place. After some time he rose from his desk, proceeded to the board, removed my name from Bowen's group and transferred it to another section with the remark that he thought it would be most sensible if I took charge of an apparatus test room to stop the waste of apparatus of which I complained.

Fortunately in a few days I moved with Bowen's group far away from Dundee and did not suffer that consequence of my intemperate intervention. Neither did I know that Rowe's apparently pleasant conversation with me on that afternoon concealed a deep irritation that complaints about one of his groups had been brought to the notice of Tizard. On 26 October he wrote to Tizard:[9]

> On my return, I visited Perth and saw Blackett's three men for the first time.
> As I have often done with men who have come in from outside, I asked them
> whether they had any suggestions to make regarding improvements in
> running the show. It was soon obvious to me that the writer of the letter
> which you quoted to me was a man named Lovell. I therefore asked Lovell
> to come over to Dundee to discuss matters with me and I have spent an hour
> with him this afternoon. He clearly has no idea that I am aware that he has
> written to Blackett. Judging purely from the letter you quoted to me I
> expected to find that Lovell was a nasty piece of work who should be
> removed from the work. I find, however, that this is not the case. Many of
> the criticisms he raises are associated with the simultaneity of research,
> development, production and installation and with the natural alterations
> of mind of C. in C. Fighter Command as he feels his way towards an
> operational solution.
>   It was rather interesting to note that Lovell said that many of the things he
> felt critical about two weeks ago he now understands. At the same time he
> had a number of weighty criticisms which boil down to the fact that Perth

has been separated from Dundee. There is no Test Room at Perth and no organisation for the ordering and inspection of materials. There are many associated detail troubles caused by the separation.

When a man comes to me with the statement that everything is all wrong, I usually start by suggesting that he should put it right. This policy has worked in this instance and Lovell has gladly accepted the job of organising a test room and taking over contract procedure at St. Athan to which Perth is moving next week. Bowen is happy with this arrangement. I hope it will work all right because I have seen little enough of Perth and am likely to see nothing of St. Athan. I feel convinced that we shall make something of Lovell after his tactless start.

That was my first real and nearly catastrophic contact with Rowe. When, in 1948 he wrote his story of the war years he remarked[10]: 'On 1 January 1942 Lovell was given charge of the H_2S project and of the sister project, centimetre ASV', but that—the main reason for this book—was separated by two years of many changes from the Scone airport and Dundee of October 1939.

References

1   Bowen E G 1987 *Radar Days* (Bristol: Adam Hilger) p 87
2   Lovell Bernard 1985 *The early history of the Anglo-Australian 150-inch telescope (AAT) Q. J. R. Astron. Soc.* **26** 393
3   Blackett archives, Royal Society D7
4   See reference 1
5   Bowen E G 1987 *Radar Days* (Bristol: Adam Hilger) p 78
6   Hanbury Brown R 1991 *Boffin* (Bristol: Adam Hilger)
7   Saward D 1984 *Bernard Lovell* (London: Robert Hale) pp 46–48
8   Lovell Bernard and Hurst D G 1988 *Wilfrid Bennett Lewis 1908–1987* in *Biog. Mem. R. Soc.* **34** 453
9   The letters Lovell to Blackett 14 October 1939; Blackett to Tizard 19 October 1939 and Rowe to Tizard 26 October 1939 are in File 32 of the Papers of Sir Henry Tizard at the Department of Documents, Imperial War Museum
10 Rowe A P 1948 *One Story of Radar* (Cambridge: Cambridge University Press) p 119

Chapter 3

St Athan

30 Oct 1934

Four days after my visit to Rowe in Dundee Bowen's group left Scone airport for St Athan, near Barry in South Wales. I arrived there at dusk on 1 November. No doubt, in the days of peace, Barry had been a pleasant seaside resort. When I arrived on that November evening it 'looked foul with the dull evening and mud in the estuary'. Eventually after walking about in a 'black out that wasn't' I found a room in a hotel.

The next morning I found the aerodrome at St Athan. The contrast of Barry and St Athan with the city of Perth and Scone was extreme. The brilliant sunshine and grass field of Scone were replaced by the murk, mud, huge hangars and concrete runways of the unfinished St Athan. No one seemed to know anything about us or to expect us and we were soon directed back to Barry to find some decent lodgings. Eventually I found a room in a guest house and shared it with Peter Ingleby until he was killed two months later.

The facilities at Scone airport were meagre but at least the place itself was agreeable. At St Athan there were no facilities whatsoever and the place was inhuman. We were allocated a vast hangar and Bowen persuaded the officer in charge to erect some canvas partitions along a part of one side to create our laboratory. Bowen has given a vivid description of the place, of attempting to work in gloves, hats and overcoats in a searing winter wind with no source of heat. 'Here was one of the most sophisticated defence developments being introduced to the Royal Air Force and it was being done under conditions which would have produced a riot in a prison farm'. [1]

St Athan was being built as one of the Air Ministry's huge training and maintenance establishments. When we arrived 4000 people were already there and every building overflowed and every facility was saturated. No one has ever established satisfactorily the procedural sequence which suddenly led to the transfer of Bowen's airborne research group to this entirely unsuitable location, not 30 miles but 300 miles from the parent research establishment which remained in Dundee. All that Bowen knew was that No 32 Maintenance Unit to which we were attached, although established primarily for airframe and engine maintenance, was now given the additional task of installing airborne radar. He was told that 'our principal job was to instruct the Unit in the fitting and testing of airborne radars, after which we could return to our proper sphere of research and development'.[2]

Very few of those of us who arrived at St Athan in November 1939 ever did return to research in the parent AMRE establishment and it is hard to understand why Rowe allowed one of his key groups to be destroyed in this way. An important factor was no doubt the serious acrimony that had developed between the Rowe–Lewis grouping and Bowen. The genesis of this was the arrival at Bawdsey of Lewis in the summer of 1939. At that stage Bowen and Wilkins were the senior and pioneer members of the research group—Bowen on airborne radar and Wilkins on the development of the main ground radar installations. However, Lewis impressed Rowe to such an extent that on the outbreak of war and the move to Dundee, Rowe 'had risked offending my old colleagues by appointing Lewis as my deputy although he had been with us but a few months'.[3] In fact, Rowe did not avoid this risk. Wilkins soon moved to Watson-Watt's London office and Bowen was forced into increasing isolation, first at Scone and then at St Athan. The concept that the maintenance unit at St Athan should be trained in the installation and testing of the airborne radars was no doubt a good idea, but to immerse Bowen's research group in this task was a disastrous idea. After only a few weeks at St Athan I wrote to Blackett (17 December) explaining what was happening—but without this hindsight as to the probable cause—arguing that the work we were doing could be carried out far more efficiently and appropriately by the engineers from the firms who were then manufacturing the airborne AI equipment.

The fallacy of the emergency move from Bawdsey and the aerodrome at Martlesham Heath was soon underlined a few weeks after our arrival at St Athan. A JU88 appeared out of the murk and dropped a 1000 pound bomb on the main runway. The story of AI radar and Bowen's group might well have been different had this bomb exploded, but it did not do so and proceeded to bounce along the runway like a tennis ball—the detonator was faulty.

As far as my own work at St Athan was concerned the see-saw continued. There were now three of us from Manchester in Bowen's group, my room-mate Peter Ingleby and R K Beattie the son of the Professor of Electrotechnics in the University. When we left Scone there were 22 Blenheims still to be fitted with the AI Mark I radar. One disappeared on the flight from Scone to St Athan and by mid-November we had bench tested the equipment for the remaining 21. Bowen[4] has given the statistics of the number of Blenheims emerging from St Athan—there were 17 in December and in 1940 they were delivered to squadrons at the rate of one a day—some with the improved engineered versions known as Mark II, III and IV. I remained marginally associated with these AI radars until the early months of 1940. The cold and the ice and snow increased and there were days when our hands were too cold to hold a screwdriver but my own life and work soon changed, first through an urgent signal from the Air Ministry and then through the tragedy that killed my two young colleagues from Manchester, Ingleby and Beattie.

Air to Surface Vessel (ASV)

In the early stages of the development of airborne radar Bowen found that the equipment gave clear echoes from ships at sea. During the Air–Sea exercises in the North Sea in early September 1937 his experimental equipment located the Fleet under conditions that had grounded Coastal Command. This episode proved to be the foundation of the development of the airborne ASV radars simultaneously with the AI systems. Coastal Command had asked for ASV to be installed in the new Lockheed Hudsons and one of these was already at Scone airport when I arrived there in late September 1939. Two types of installation had been developed; one was for homing, with two dipoles with reflectors, to give overlapping beams forward of the aircraft so that the pilot could be given left–right directions similar to the azimuth direction system of the AI. The other configuration, used for search, employed directional Yagi aerials mounted to port and starboard of the fuselage. These produced a narrow beam and consequently gave greater range than the homing configuration. Otherwise the transmitters, receivers and display arrangements were fundamentally those of the AI radar.

When we arrived at St Athan the first production models of ASV Mark I were just being delivered and on 4 December Bowen received a signal from the Air Ministry ordering him to give top priority to the installation of these ASV equipments in the several Hudson aircraft now at St Athan, and in the Sunderland flying boats at Pembroke Dock. Overnight I found that my task had once more been switched from the AI radar to the priority ASV installations in the Hudson aircraft.

The reason for this abrupt change of priorities appears to have been the failure to detect the return of one of the German pocket battleships, the *Deutschland*, to Germany. The *Deutschland* and another pocket battleship the *Admiral Graf Spee* were already through the danger zones of detection when war began. The *Deutschland* had orders to harass our lifelines in the North Atlantic and did so with an efficiency that caused great concern to the Admiralty, to the extent that twelve powerful ships and three aircraft carriers from the Fleet were diverted to protect the convoy routes.[5] Early in November the *Deutschland* returned to Germany† without being detected by the Navy, Fleet Air Arm or Coastal Command. The natural presumption was that if sufficient Coastal Command aircraft had been operating with ASV the *Deutschland* could not have avoided detection.

† The *Admiral Graf Spee* was scuttled on 17 December 1939, off Montevideo after suffering severe damage from British cruisers in the battle of the River Plate.

Consequently the pressure for the installation of the ASV radars became great and on 19 December Bowen was ordered to suspend all further installations of AI until 1 January. By the end of the year the installation in three of the Hudsons had been completed. Compared with the Blenheim night fighters the Hudsons were spacious and on 21 December, on a test flight over the Bristol Channel, I recorded that the ASV performance was magnificent. That was a typical day of that period at St Athan when, after test flying, we would work until midnight, when we would be frozen beyond the limits of activity, in order to get the equipment working. On that day, for example, I noted that 30 receivers had been delivered from the manufacturers (Pye) but that they were 'absolute junk' and we only managed to salvage six suitable for installation in the Hudsons.

The AI and ASV fitting programme continued in this vein through the early months of 1940, but my own life and activity soon underwent a fundamental change. On 7 January Ingleby and Beattie, my two young colleagues from Manchester, were flight testing another ASV installation when the Hudson crashed into a 1500 foot hill near Bridgend. There was fog and, together with the flight crew and three other service men under training in the maintenance of the radar, they were killed. I was distraught, so far the war had meant merely discomfort; suddenly the young colleague with whom I had shared a room had suffered a violent death. Soon we were to become accustomed to the violence of war on our doorsteps but that time still lay in the future. Then 'my mind was like a dull ache, reconstructing everything connected with it. St Athan was like hell itself with a cold clammy hand of fog settling in all over the foul place'.

Hope soon emerged from this tragedy. The dignified bearing of Ingleby's father as he came to collect Peter's belongings was one lesson I still had to learn. Joyce came from Bath, found a house to rent, brought down the one-year old Susan and made a home where all water pipes remained frozen for two months. After a few days Blackett came to stay with us—we walked along the cliffs overlooking the channel in brilliant moonlight and began to talk about the radar echoes of unknown origin which, years later, were to lead me to Jodrell Bank. But on that night Blackett was more interested in the ships in the channel and found that he, as an ex-naval officer, could not compete with Joyce, an ex-Girl Guide, in the reading of the morse code flashing from the ships. That was 17 January but weeks of that arctic weather lay ahead in which rumour succeeded rumour as night follows day about the future of Bowen's group.

The Emergence of Research

That January visit of Blackett's was not the first time he came to St Athan. Disturbed by my letters describing the misuse of scientists as maintenance fitters he had visited St Athan on 14 December with Tizard and Watson-Watt (then DCD—the Director of Communications Development) to see what was happening to the AI and ASV. They were greatly disturbed. By mid-January, when Blackett came again, he said that part of Bowen's group would be moved and re-created as a research team, leaving Touch and some others at St Athan to look after the installation and maintenance work on AI and ASV. The alternating rumours centred mainly around a return to Dundee, to GEC in London or to RAE in Farnborough, but these bore little relation to the actual happenings as the snow and ice of that fearful winter disappeared. Meanwhile, King George VI and Queen Elizabeth braved the rigours of the winter to visit St Athan on 9 February. We were lined up in the bitterly cold hangar and as Bowen later wrote: 'He passed down the line and had a pleasant word for every member of the group. We were honoured by the visit and it gave an enormous boost to a group which had been feeling the cold, cold winds of adversity in more senses than one'.[6]

On 7 January 1940, when Ingleby and Beattie were killed in the crash of the Hudson aircraft, I not only lost two colleagues but also the only two university members of the group who had worked with me on whatever small amount of research had been possible at Scone. A H Chapman, Blackett's cosmic ray technician, had joined us. In the few weeks before leaving Scone Bowen had asked me to begin some research work on the project then known as BD. The plan was to detect by radar, from above, an incoming enemy force, and drop bombs which would be exploded automatically by a radar impulse from our high flying airplane. The four of us had devised a system which we thought worth trying in flight and had this fitted in an Anson ready for flight trials when the move to St Athan took place. I never saw that Anson again, and that was the last piece of anything resembling research that I was to do with Ingleby and Beattie.

One other university person had joined Bowen's group—John Pringle, a young zoologist from Cambridge and although he lived with us for some time after Joyce had made a home in Barry, we saw little of him since he had been diverted to Pembroke Dock to deal with the ASV installations in the Sunderland flying boats.

When Blackett came on 17 January conveying the hope of re-creating a research team he realised immediately that one man alone had only the slenderest hope of making any useful progress where all the priority and men were on the fitting and testing of the existing AI and ASV. It was during that visit that he promised to transfer an 'alpha plus' man to join me at St Athan. In that manner Alan Hodgkin arrived at St Athan on 26 February. He had nowhere to stay the night so I took him to our home in Barry and in that way the kind of life that I thought had vanished forever began to re-appear. Alan Hodgkin's name will appear frequently in the pages that follow. Blackett's judgment of the young man was infallible. In 1970 Hodgkin was to succeed Blackett as President of the Royal Society, and to become a Nobel Prize Winner and Master of Trinity.

None of that was even a dream on that February day in 1940 when the young physiologist arrived at St Athan. At last we were allowed to begin a little research. The two of us with Chapman's help made a large horn-type antenna and on 5 March we took it outside the hangar and fed it with a 50 centimetre oscillator. We visited GEC at Wembley where M R Gavin showed us a new design of valve capable of generating several kilowatts of peak power in pulsed operation. This was known as the 'micropup' (and officially as the VT90). This valve, of radically new design, consisted of an outer cylindrical anode with fins attached for air cooling. The anode, made of copper, was part of the envelope of the valve and attached to each end was a short sealed length of glass tubing. The grid was at one end on a glass–metal seal and the cathode was at the other end.[7] Alan Hodgkin and I were fascinated by this new development and for the first time the practical possibility of an AI on a wavelength of 50 centimetres with a narrow beam to avoid ground returns began to emerge. We returned to St Athan and our very large horn. Even if it were possible to fit such a device in a night fighter some means had to be found, other than by moving the horn, of swinging the narrow beam. At least facing a seemingly insuperable research problem with a fellow spirit was a refreshing change from the preceding months of drudgery with the AI and ASV equipment.

We continued with this work on horns, measuring the shape of the beam, and trying to devise a means of swinging the beam with the horn stationary, against the uneasy background of constant discussion and manoeuvres about our future. Touch spent more and more time at the manufacturers of the equipment and in February he moved his office to RAE Farnborough. This was the first significant move in the disintegration of Bowen's group and, incidentally, left us with the freedom to proceed with the research on the shorter wavelengths. On 29 March a visitor from DCD's headquarters told us that Bowen's group had already been officially 'dispersed' and in the first days of April we were told that the research part of the group—mainly Hodgkin and myself, with the horns—would re-join the main AMRE establishment which was to move from Dundee to 'somewhere' on the Dorset coast near Swanage.

Bowen recollected St Athan as 'the place where we performed an enormously complicated juggling act—which was far more complex than anything ever seen in a three-ring circus … done with totally inadequate resources and virtually no administrative back-up'. Nevertheless he goes on to say that the main task of handing over the installation and maintenance of the 1½ metre AI and ASV radars to the RAF was 'enormously successful. Within a few months of our arrival St Athan was fitting one aircraft a day with either air-interception or sea-search radar and, as time went on, these figures were far exceeded'.[8] In the early days of April this task was left altogether to the RAF unit at St Athan. Bowen's original airborne group dispersed, some to RAE, others to assist in the maintenance of the airborne radars on the operational stations, and the fortunate few of us from the universities re-joined the main AMRE establishment in process of transferring to the Dorset coast.

In the pioneering work on the airborne radars and in the introduction of the AI and ASV Mark I radars to the RAF operational commands Bowen had carried through a remarkable task. Unfortunately, for reasons already described his relations with Rowe and Lewis steadily worsened. After the dispersal from St Athan new developments of great significance to AI and ASV took place. For a few months he remained on the fringe of these activities but never became integrated into the core of the team of new arrivals. Early in September 1940 he crossed the Atlantic by sea as part of the Tizard mission carrying in a black box one of the first cavity magnetrons, later described by an American† as 'the most valuable cargo ever to reach our shores'. Apart from occasional visits he did not return to England and it was the Americans who henceforth benefited from his unique experience.

References

1   Bowen E G 1987 *Radar Days* (Bristol: Adam Hilger) p 93
2   Bowen E G 1987 *Radar Days* (Bristol: Adam Hilger) p 92
3   Rowe A P 1948 *One Story of Radar* (Cambridge: Cambridge University Press) p 57
4   Bowen E G 1987 *Radar Days* (Bristol: Adam Hilger) p 95 and appendix I
5   Churchill Winston S 1948 *The Second World War* vol 1 The Gathering Storm (London: Cassell) Chapter XXIX
6   Bowen E G 1987 *Radar Days* (Bristol: Adam Hilger) p 97
7   A detailed account of the evolution and characteristics of this type of valve has been given by Bell J Gavin M R James E G and Warren G W 1946 *J. Inst. Elect. Eng.* **93** pt IIIA 833
8   Bowen E G 1987 *Radar Days* (Bristol: Adam Hilger) p 135
9   Wilkins M H F 1987 *John Turton Randall in Biog. Mem. R. Soc.* **33** 493 in section *The Cavity Magnetron* pp 502–510
10 Watson-Watt Sir Robert 1957 *Three Steps to Victory* (London: Odhams) p 297
11 Bowen E G 1987 *Radar Days* (Bristol: Adam Hilger) p 194
12 Baxter J P 1946 *Scientists Against Time* (Boston: Little, Brown)

† In his biographical memoir on J T Randall, M H F Wilkins[9] ascribes this comment to President Roosevelt. However, Watson-Watt[10] and Bowen[11] state that the originator of the comment was the official historian of the US Office of Scientific Research and Development (OSRD), James Phinney Baxter III[12] who wrote (of the magnetron): 'When the members of the Tizard Mission brought one to America in 1940 they carried the most valuable cargo ever brought to our shores. It sparked the whole development of microwave radar and constituted the most important item in reverse Lease–Lend'.

Chapter 4

Worth Matravers

John Pringle spent most of his time at Pembroke Dock on the fitting and flight-testing of ASV in Sunderland flying boats but after Touch moved to Farnborough in February, he was the senior member of the scientific officer class in Bowen's group and during Bowen's several long absences he assumed administrative control. It was with him that I had so many discussions about the future of the group and ourselves. At the end of March 1940 when it became clear that our days at St Athan were numbered we decided that John Pringle should visit Rowe in Dundee with the hope that he could clarify the future of the group, or at least our part of it. On 8 April he returned with the news that at least the recruits from the universities were to rejoin the main AMRE establishment which was moving to the Dorset coast near Swanage early in May. The next day, as Hitler's troops invaded Denmark and Norway, it was decided that I should go immediately to organise the move and especially to find the airfield at Christchurch which we were to use for our airborne work.

Two days later I was at Worth Matravers. It was a spring day of brilliant sunshine and, with the sea sparkling below the cliffs, the entire scene was exotic. The chosen site for AMRE was a little beyond the village of Worth Matravers. The shells of several buildings were erected and Joe Airey, to whom I had been told to report, was confident that the establishment from Dundee would move in on 4 May. Airey had been head of the workshops at the Radio Research Board at Slough and had moved with the pioneer group to Orfordness in 1935 and then to Bawdsey and Dundee. On that day the prospect of working at Worth Matravers appeared most agreeable. There were no facilities for an experimental establishment but no doubt they would materialise and so I set off with Airey to find the aerodrome at Christchurch. With the vast hangars and runways of St Athan in mind I was surprised to find no sign of any similar establishment. Eventually I decided to ask a local shopkeeper, a tobacconist, for directions. He was puzzled and could think only that I was referring to the small local flying club. He was correct. Within the walls of a London office one airfield was presumably like any other airfield. The concrete runways of St Athan had replaced the green fields of Scone, and, in turn, St Athan was to be replaced by an even smaller

field than the one at Scone. At least Scone had an officer in charge and
two small hangars. Christchurch had a wooden hut containing a couple of
light aircraft, a cafe (closed) and a flying club room. The private houses
surrounding the field made any prospect of extension out of the question
and I could not imagine how our experimental aircraft could operate with
safety in that environment.

A few days later I returned to St Athan and telephoned DCD to tell him
so. In January Sir George Lee had replaced Watson-Watt as DCD and I had
already had a pleasant conversation with him when he visited St Athan
towards the end of that month. Now I was to learn that a Junior Scientific
Officer was not expected to telephone the Director, particularly if it was to
convey news of an unsatisfactory nature. Sir George was abrupt to the extent
of rudeness—his staff could not possibly make such a mistake. But they had
done so and in the event we had no option but to use that bumpy field, no
doubt satisfactory for a Tiger Moth but not for a Blenheim night fighter.
Worth Matravers was a delightful place in which to work but, for those of
us who had to fly, that airfield at Christchurch was a grim reminder of the
discomforts and dangers of wartime existence. About one and a half years
later the major aerodrome at Hurn was operational and we transferred our
airborne work there, but, as will be related, that was a very short-lived
arrangement because in the spring of 1942 we were forced to make yet
another move. In fact, by far the major part of our airborne work during
the next 18 months was carried out from Christchurch, using the small
clubhouse as a makeshift laboratory.

On 4 May we left St Athan—and our temporary home in Barry—without
any feelings of sorrow. In the few months on that aerodrome we had
accumulated a large amount of equipment and, as the organiser of that
move, I had loaded three 3-ton and one 1-ton articulated trailers as well as
a series of van-loads. The arrival of these loads at Worth Matravers on 7 May
caused consternation. We had pre-empted the arrival of far less efficient
transports from Dundee. Rowe had issued an order that no one was to leave
until these had arrived and were unpacked and that meant wasting five hours
after our own transport had departed. Nevertheless for a few of us, at least,
a very different and exciting phase of our wartime research had begun.

The First Centimetric Radar

Immediately after our arrival at Worth Matravers a new team began to form,
concerned solely with the problem of developing an AI radar on a short
wavelength so that ground returns could be avoided. Herbert Skinner, who
had been one of the Bristol lecturers during my undergraduate days, had
spent the winter installing a coastal defence radar in the north of Scotland

and he came with the main AMRE establishment and so did J R Atkinson. On 15 May Skinner and I drove to Bristol to collect a magnet and a split-anode magnetron so that we could generate small power on a wavelength of 10 centimetres for measuring polar diagrams. On that day P I Dee arrived and suddenly we became a group of research physicists again.

The arrival of Dee was an important event in the evolution of Rowe's establishment. He was a distinguished physicist in the Cavendish laboratory where he had been senior to Lewis. Since Lewis was now Rowe's deputy this fact alone had a major effect on the influences in the establishment. In the immediate pre-war period, because of the nature of their researches, it was Lewis who had the greater familiarity with radio techniques. Primarily for this reason, just before the outbreak of war Dee had chosen to join the Royal Aircraft Establishment (RAE). He had taken with him from the Cavendish a group of young physicists including W E Burcham, C W Gilbert, J G Wilson, S Devons and S C Curran (who was soon to marry Jane Strothers, another member of that group). With this group Dee had gone to the Air Defence Department of the RAE which was evacuated to Exeter. There, in the early months of the war, Dee had conceived and developed the parachute and cable (PAC) rocket defence weapon for use against low-flying aircraft. Early in April 1940 Dee was asked to report to Rowe's establishment which was about to move from Dundee to Worth Matravers. This he did with Burcham and Wilson on 15 May. According to Burcham's diary[1] they were 'given a piece of paper at the gate which permitted them to see Lewis, we were shown round the Establishment by him. Our tour included a visit to "a hut in a field where they had big horns and mirrors for canalising radiation"'. Subsequently at various stages the other members of Dee's RAE group arrived.

The hut to which Burcham refers was in a field (known as C site) even closer to the cliffs than the main part of the establishment, but for how long we would be allowed to continue working on ideas that were very far away from any kind of operational use suddenly became doubtful.

On 10 May Hitler's armies had invaded Holland and Belgium. Chamberlain resigned and Churchill replaced him as Prime Minister. Brussels, Antwerp and many of the aerodromes in France were bombed. Our Expeditionary Force retreated and was cut off from the French armies as the Germans surged westwards. All leave was cancelled and Rowe soon came under great pressure to devote the whole of his establishment to the radar devices of immediately operational significance. Indeed, it required an act of great faith to believe that any radar device not then developed would ever be used in the war. In 1988 when I was writing the biographical memoir of Lewis I discovered that it was he who had that faith and with great difficulty had persuaded Rowe to allow the mixed group of university recruits to continue their work on a futuristic AI.

The Problem of Generating High Frequencies

The frequency at which a triode valve operates is governed by the inductance and capacity of the associated circuits. By merely changing the external circuit parameters, a conventional triode can operate at frequencies up to about 30 megahertz without serious loss of efficiency. At higher frequencies the efficiency falls rapidly because the time period of the oscillation becomes comparable with the transit time of the electrons from filament to grid and anode in the valve, and hence the phase relations are disturbed. To a certain extent the loss of efficiency can be compensated by using two triodes in some form of push–pull circuit and it was that form of transmitter, employing two triodes in a squegging oscillator, that was in use in the first AI which was being installed when I arrived at Scone in September 1939. That transmitter operated on a frequency of 200 megahertz (1½ metre wavelength) and that represented the upper frequency limit at which the conventional triodes could be used with reasonable efficiency.

In the 1930s, the development of television, to a certain extent, stimulated the investigation into the design of thermionic valves which could operate efficiently at high frequencies. Miniaturisation to reduce the transit time of the electrons inside the valve was an obvious approach. Already in 1933 in America, the small dimension 'acorn tubes' of RCA and the 'doorknob' tubes of Western Electric were on the market. These had a short electron transit time and were designed to have a small interelectrode capacity with connections to external circuits of low capacity and inductance. Using miniaturised Western Electric tubes of this kind Bowen had a ground radar system working on a wavelength of 30 centimetres in 1938 but the power was limited and the range of detection too small in comparison with the 1½ metre systems to justify airborne trials.

Late in 1939, while we were still at St Athan, the Air Ministry had placed contracts with the GEC Research Laboratories in Wembley and with EMI at Hayes for the development of a short wavelength AI radar. Bowen had acted as coordinator of these meetings. This had stimulated the GEC research into methods of producing adequate powers on centimetre wavelengths, and the micropup (VT90), which I was shown when I went to GEC with Alan Hodgkin in March, was one result of this work. The micropup represented an advanced form of the development of the conventional type of triode in which the electron transit time had been reduced by severe reduction of the spacing between the electrodes and the circuit constants of the tube itself had received particular attention—in the original micropup design for example, the anode was cylindrical and formed part of the envelope of the valve. In these developments the valve was treated as an integral part of the overall circuit. A pair of micropups could give 10 kilowatts of pulsed power at a wavelength of a metre and later GEC developments of this type could yield several kilowatts of pulsed power on a wavelength of 25 centimetres. By the summer of 1940 GEC had a complete radar working on a wavelength of 25 centimetres installed on the roof of the Wembley laboratory. This gave some promise that an operational system of this type could result from further development. However, this line of development using conventional, though specialised, thermionic valves almost immediately gave place to remarkable developments elsewhere.

The Klystron

The Committee for the Coordination of Valve Development (CVD) covered the interests of the three armed services and, in the early months of the war, this committee had placed a number of contracts for the development of transmitting and receiving valves on a wavelength of 10 centimetres. One group worked at the University of Bristol—and this was the reason for my visit there with Skinner soon after we arrived at Worth Matravers. Another contract was placed with Professor M L E Oliphant's group at the University of Birmingham and it was from there that a revolutionary development occurred which transformed the outlook for centimetric radar.

The initial work of the group at Birmingham was on the klystron. In 1936 in the USA interest developed in the properties of resonant cavities—that is, a region bounded by an electrically-conducting shell—such as a sphere or circular cylinder made of copper. W W Hansen[2] at Stanford had investigated the properties of the cavity resonator. He stated that interest in these resonators arose because of the possible application to produce high-speed electrons. A stream of electrons was projected across the resonant cavity by an intense alternating field produced by a high frequency thermionic valve. Electromagnets reversed the direction of the electrons every half-cycle so that a number of transits of the electrons through the cavity could be achieved before they were allowed to escape. This device was known as a rhumbatron. Although the application to the production of high-speed electrons did not develop until the post-war years, the principle rapidly led to the invention of the klystron involving the combination of two rhumbatron cavities into one tube.[3]

In the klystron a beam of electrons is projected through two adjacent rhumbatron cavities (figure 4.1). The high frequency field is applied to the first rhumbatron cavity (the buncher) which bunches the electrons by alternately accelerating and retarding them. The fast and slow electrons are separated in their passage through the field-free space before they enter the second cavity. That is, the original electron stream has been intensity-modulated by the buncher. When this modulated electron stream passes into the second cavity (the catcher) it excites the resonator which is tuned to the appropriate frequency. Power is taken from the catcher by a loop or other device. In the case described above the buncher is driven by an external power source—such as the signal from an antenna and the klystron is acting as an amplifier. In another mode, if the buncher is driven by power received through the coupling loop from the catcher, then the klystron will act as an oscillator. This idea of combining two resonant cavities to act as an amplifier, or oscillator, for very high frequencies originated in discussions between Hansen and Varian in the spring of 1937 and by the end of the year two tubes, continuously evacuated, were giving energy at a wavelength of 12 centimetres.[4] Oliphant's interest in this device which he saw during a visit to the USA in 1938 was to have far-reaching consequences.

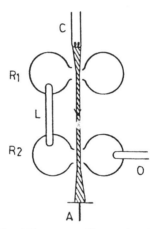

Figure 4.1 Principle of klystron oscillator. An electron beam from the cathode C is focused through two cavity resonators (rhumbatrons) R1, R2 on its way to the anode A. The beam is velocity-modulated in R1 and builds up a similar oscillation in R2 which is fed back to R1 by the loop L. Power is extracted by the coupling loop O. (From *Fifty Years of the Cavity Magnetron* by W E Burcham and E D R Shearman reproduced by kind permission of the University of Birmingham.)

The Magnetron

Initially the work of Oliphant's group was concerned with the development of an improved klystron oscillator. By the end of 1939 they had built a klystron which produced about 400 watts of continuous wave power on a wavelength of 10 centimetres. During the next few months they produced a klystron which could be pulsed and also serve as a local oscillator.†

† I am indebted to W E Burcham for the following comment on this development: 'the idea was to produce a strong transmitter pulse at one frequency ν_1, and then, after a small delay, a voltage-tuned, weaker pulse at a frequency $\nu_1 \pm 45$ Mc/s which would be ready to mix with the reflected signal ν_1 in a heterodyne system using a crystal detector. Unfortunately it was all too clever and I do not believe that it was ever seriously used even before the Sutton tube.' (see p 61–3)

However, the problem of producing a klystron to fit in an airborne equipment capable of yielding the many kilowatts of pulsed power needed remained severe. A major problem was that of producing an electron beam of significant intensity in a sealed-off tube to be airborne.

In the event, although small klystrons were to be widely used as local oscillators in airborne radar, the solution for the production of high-pulsed powers came from an inspired idea of J T Randall and H A H Boot which led to the resonant cavity magnetron. The simple form of magnetron had a long pre-war history. In the 1920s, both in the USA and in Japan, simple magnetrons had been built which produced small powers on a wavelength as short as 10 centimetres. These consisted of a cylindrical anode with a straight wire cathode along its axis. A magnetic field was applied along the direction of the cathode so that the emitted electrons pursued a curved path without reaching the anode. The oscillations produced depended on the time taken by the electrons to describe the curved path. An important development was described in the mid 1920s when it was discovered that if the cylindrical anode is split into two half cylinders the device exhibits negative resistance properties and could maintain oscillations when connected to a tuned circuit. It was this simple form of split anode magnetron which I obtained from GEC and used to measure the polar diagram of the various aerials at Worth Matravers in the spring of 1940.

Subsequently a considerable literature developed on the use of this type of magnetron to produce useful powers at high frequencies[5]. However the idea of Randall and Boot does not seem to have had any relation to these pre-war developments of the simple magnetron. Oliphant had asked them to investigate the possibilities of using the Barkhausen–Kurz effect as a receiver in the 10 centimetre wavelength region. In 1920 Barkhausen and Kurz applied a high positive potential to the grid of a triode and a small negative potential to the anode so that electrons were accelerated through the grid but were returned to the filament without reaching the anode. They found that oscillations were produced at a frequency governed by the time taken by the electrons to travel from the filament, through the grid and back to the filament again. In the autumn of 1939 Randall and Boot wanted an oscillator on 10 centimetres to test their Barkhausen–Kurz tubes as receivers and turned their attention to the principle of the magnetron for this purpose. The sequence was described by Randall in an address to the Physical Society of London on 12 December 1945 after he was presented with the Duddell medal of the Society.[6]

> In the early autumn of 1939, with others whom Professor Oliphant had collected together for this purpose, I began to spend some time on problems connected with centimetre waves, as many other physicists were doing elsewhere. The chief problem was to produce a high-power centimetre wave oscillator. The centimetric nature of the wave was required for direction of beam, and the power for long range. At first Boot and myself spent a few weeks studying the Barkhausen–Kurz tube as a detector device; the greater

number of workers in the laboratory, however, were concerned with the klystron, both as an oscillator and as an amplifier, and we were naturally interested in the whole field. The klystron was outstandingly important in that it used for the first time closed, or essentially closed, resonators, fitted with grids through which an electron beam passed, for the production of high-frequency power. Such resonators had been considered by Rayleigh in 1897, and the papers of Hansen showed how enclosures of this kind lend themselves to certain modes of electrical vibrations. On any given mode the frequency is largely, if not entirely, dependent on the dimensions of the resonator. When such resonators are constructed of copper, three important features are evident: (i) low h.f. losses; (ii) wave-length stability; (iii) potential capability of large heat dissipation.

To be set against this, however, is the serious difficulty of getting sufficient power to the electron beam, which is necessarily of small cross-sectional area. We began, therefore, to think of other possibilities. Was there any means of using the desirable features of cavity resonators, and at the same time avoiding the limitations of the electron beam?

Conventional designs of magnetrons were equally notable in failing to meet the desirable features, and by November 1939 we had decided to try to devise a type of magnetron which did so. The Hansen papers already referred to were by this time available in the laboratory, but it was obvious that the types of resonator considered—hollow spheres, cubes, etc.—could not be associated with the cylindrical anode and cathode of the magnetron. The only other types of short-wave resonator circuit with which we were familiar were Hertz's original loop-wire resonator and a short-circuited quarter-wave line. A three-dimensional version of the first of these gives a cylinder with a slot down one side, and the second becomes a parallel-sided slot $\lambda/4$ deep. Either design was well suited to the arrangement of a number of segments with cylindrical symmetry round a central cathode, and it is also clear that a combined anode and resonator system could be machined from solid copper to allow good heat transfer and, it was hoped, high-power dissipation.

No theoretical calculations of the wave-length to be expected from a resonator of the cylinder-slot type had been made, but Hertz had shown, and Macdonald had confirmed theoretically, that the wave-length to be expected from a simple Hertzian dipole was $\lambda = 7.94d$, where d is the diameter of the ring.

Six resonators, 1.2 cm. in diameter, with slots 0.1×0.1 cm., were employed, and the length of the resonator was 4 cm.; it was hoped that the wave-length of this system symmetrically surrounding an anode-hole of the same diameter would be approximately 10 cm. Preliminary calculations had shown that this design might be expected to operate reasonably well with an applied potential of 16,000 volts and a magnetic field of approximately 1000 oersteds, so that provision of these essentials was a necessary preliminary. In order to save time a D.C. power-supply system was built up, and not a modulated supply, and a large electromagnet was used to give the required field parallel to the cylindrical axis of the valve. The first resonator magnetron was continuously operated on the pump with glass-metal junctions closed with sealing wax. For various reasons the first trials were delayed, and it was not until 21 February 1940 that the valve was tested. It worked immediately. By various rather crude means it was shown that the output was about 400 W. The wave-length was measured by means of Lecher wires and shown to be 9.8 cm.

Figure 4.2 This cavity magnetron of Randall and Boot first produced power on 21 February 1940. (University of Birmingham)

In spite of this dramatic success, considerable doubts remained as to whether the magnetron could be used in radar—principally because magnetrons were notoriously unstable in frequency. An important modification was introduced by J Sayers in August 1941, who discovered that by connecting alternate cavities with a 'strap' (a copper wire) the oscillations could be kept in one mode. When it was realised that the resonant cavities then exerted a dominant influence on the frequency control, these prejudices disappeared. In March 1942 at TRE Devons[7] showed that the magnetron frequency could be pre-set during manufacture by adjustment of the straps and, with this further aid to mass production, the magnetron superseded the klystron as a high power generator of centimetre waves—but, as will be related, the klystron–magnetron conflict was to surface again under critical conditions in 1942.

In April 1940, in great secrecy, liaison was established between the Birmingham group and E C S Megaw of the GEC research laboratories with a view to the design of a sealed-off version of the Randall–Boot cavity magnetron.[8] Megaw made significant contributions to the transformation of the laboratory model of the cavity magnetron into a version that could be manufactured. Two important technical advances were involved in this transformation. In this E C S Megaw was assisted by S M Duke†, who worked both at GEC and Birmingham. He developed the gold-seal technique in which the vacuum seals involving copper-to-copper were compression-bonded by gold wire rings. A further important change was the replacement of the tungsten filament cathode by the oxide-coated cathode developed in Paris by H Gutton.[10] Two sealed off magnetrons were produced by GEC by June 1940. With the anode at 10 kV and in a magnetic field of 1500 gauss these early magnetrons could produce a pulsed output of 10 kilowatts. The prospects of AI radar and many other forms of ground and airborne radar were immediately transformed by this remarkable development.

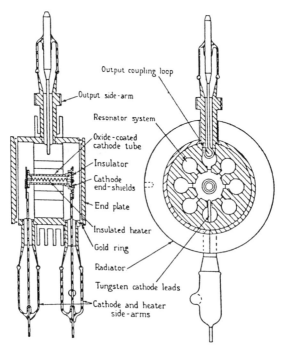

Output coupling loop

Output side-arm

Resonator system

Oxide-coated cathode tube

Insulator

Cathode end-shields

End plate

Insulated heater

Gold ring

Radiator

Tungsten cathode leads

Cathode and heater side-arms

Figure 4.3 Internal construction of the cavity magnetron as used in the first airborne centimetre AI. (From *Metres to Microwaves* by E B Callick, published by the I.E.E., reproduced by kind permission of the author and Peter Peregrinus.)

The Ground Trials at Worth Matravers

In May 1940, after our arrival at Worth Matravers, I erected the horns outside the hut that we had been allocated and soon had the split-anode

† For more details of the work of Randall and Boot and the important contribution made by Duke and others in the University of Birmingham, see the biographical memoir of Randall by M H F Wilkins.[9]

magnetron generating enough power on 10 centimetres to continue with the polar diagram measurements. At that time I knew nothing of the dramatic developments in Birmingham. The security surrounding that work was intense and it was later believed that only Rowe and Lewis in AMRE had been told of the Randall–Boot successes. † We knew, of course, of the possibilities for the klystron and, at least for some time, it was with that in mind that the new group worked on the development of a centimetric airborne radar. In his personal diary[12] written about the day of his arrival at AMRE in mid-May, Dee noted that:

> Oliphant at Birmingham is said to have developed a klystron giving a kw at 8 cm and Randall also in Birmingham has a magnetron on the same wavelength. There are no transmitting valves at all in AMRE on such wavelengths and Skinner and I have decided that I should go to Cambridge, and try to get a klystron made in the Cavendish to Oliphant's design as quickly as possible.

On 22 May, Dee visited Birmingham and wrote that he 'saw Oliphant's klystron and magnetron work', but he gave no indication that the magnetron was of a revolutionary design. On his return to Worth Matravers he recorded that he spent the following days with W E Burcham (who had arrived with him) installing the pumps and high voltage equipment for the klystron. However, it was several weeks before they had a klystron working and on 5 June he wrote:

> Lovell has got a low power 10 cm glass envelope magnetron working in a 'field' coffin for the polar diagram measurements... much to Lovell's annoyance everyone twiddles the knobs on every possible occasion, this being the first working apparatus.

However, I was soon to suffer less, because on 13 June, the continuously-pumped klystron in the hut began to produce considerable power in the centimetre waveband, but Skinner irritated Dee because his method of testing the power was by lighting his cigarette from the output lead. Indeed, Dee's account of 13 June 1940 gives a fair impression of the ramshackle nature of the research at that time. Skinner was continually burning out the filament of the klystron and this meant re-fixing and outgassing, but:

> Whenever I am out of the lab. and Skinner has to do this he forgets to turn off the water before pulling off the cooling pipes with the result that I am standing all day in about ½" depth of water, and the water on the bench is about equally deep but has its surface relieved somewhat by floating cig-ends, tea leaves, banana skins, etc. However, we have managed to get quite a lot of power out of it and light a Pea lamp nearly a foot away.

† E B Callick[11] states that at the CVD meeting on 5 April 1940, W B Lewis described the Birmingham magnetron and the results obtained on the pump.

Burcham's diary[13] gives a similar impression of the extraordinary nature of the research in that hut on C site (known as hut 40) and of the relations between Dee and Skinner, both accustomed to be in control of their own laboratory. On 20 June he wrote:

> Today Skinner was back from Bristol having smashed the magnetron at Bristol again. I retired somewhat from operations around the klystron today and there were numerous passages between D and S of a rather amusing nature. Each would like to have the klystron to himself. Atkinson flits in and out and I try to get a look-in when I can.

and his diary entry of 16 July epitomises the somewhat arbitrary and roaming brilliance of Skinner. Lewis had suggested to Burcham when he arrived at Worth that he should develop a crystal mixer:

> 16 July 1940 In the lab I started to do some glass blowing on crystals but didn't get very far before Skinner came along and said what about pulsing the klystron. So I tried that and immediately Skinner sat down at the table and started to do the glass-blowing. By the end of the day he had produced a rather nice crystal (detector) which rather peeved me, as I had wanted to do this.

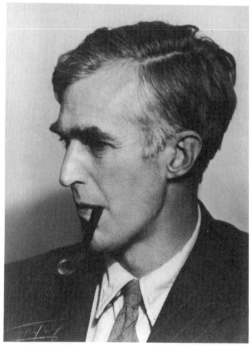

H W B Skinner photographed in 1945. (Courtesy of Elaine Wheatley, daughter of the late H W B Skinner.)

My task was to concentrate on the aerial system and also by making absorption measures to find a suitable material from which the nose of the Blenheim night fighter could be made. On 14 June Dee wrote:

> Lovell is now measuring polar diagrams on 10 cm in the field outside the hut. His mirrors provide excellent targets for competition between Atkinson and myself of the throwing of large lumps of mud which collect outside the hut.

The reference to 'mirrors' in Dee's diary note is to the parabolas which had replaced the horns. Although the horns gave good polar diagrams they were more than a yard long and it seemed impossible that aerials of that kind could ever be used in a night fighter. In the early days of June, Skinner had persuaded me to try a cylindrical (sectional) parabola† and on the eleventh I measured the polar diagram. The parabola was only 22 centimetres deep and, to my amazement, the polar diagram was the same as the horn of more than ten times that depth. Today when paraboloids are such a common feature of everyday life it is hard to believe that in June 1940 so little was known about these matters. At that time we were working in a largely unexplored wavelength region. Much later we learnt that the properties of paraboloids as aerial systems for the transmission and reception of very high frequency waves had been explored in other countries but at that time, in 1940, we knew nothing of that work and proceeded to rediscover these properties for ourselves. Indeed, on 12 June, I moved the dipole across the mouth of the parabola and found that it shifted the beam direction by 8 degrees for 5 centimetre movement. This seemed an enormous step forward in producing a narrow beam from an aerial which one could imagine fitting in a night fighter aircraft—and furthermore the possibility of moving the beam by displacing the dipole at the focus seemed to be such an exciting discovery that I wrote: 'This makes me regard the aerial problem as 75% solved'.

In the meantime the German armies had pinned the BEF against a narrow strip of the French coast and, as the heroic Dunkirk evacuation proceeded and Churchill said that we had suffered a 'colossal military disaster', it was hard for us to believe that any of the work we were doing could ever be of the slightest use in the war. However, as the Germans entered Paris and France surrendered we continued with our work. Atkinson managed to convert my CW magnetron to pulsed operation and on 19 June I measured the polar diagram of a 6 foot sectional parabola. Life became a mixture of

† J R Atkinson informs me (private communication 10 January 1991) that Skinner had arranged for this parabola to be made in March 1940 while they were still in Dundee. According to Atkinson this was the result of a letter from Sir George Lee (DCD) to Rowe instructing him to initiate work on 10 centimetres although as Atkinson writes this was 'before we had any idea of how we could either generate or receive 10 centimetres'.

normality and preparation for disaster. Churchill warned the nation that an invasion was probable and as the midsummer sun rose over the cliffs of our field site we were issued with packing cases 'in case we evacuate, but no one wants to'. In the cool of the evening we played tennis and as the sirens wailed in the middle of the night we would crouch in the well of the stairs —pronounced to be the safe refuge in a collapsing house.

In this atmosphere on 2 July Oliphant and Sayers arrived from Birmingham bringing with them a complete centimetre klystron system. † With this (operating as a continuous wave transmitter) we began investigating the signals scattered from objects on this wavelength because the defeatists, who said 10 centimetre AI would never work, argued that the reflection from a target aircraft would be so specular that this wavelength would be useless in operation. They were soon to be proved wrong in a memorable experiment.

The measurement of the polar diagrams of the sectional (cylindrical) parabolas had left little doubt that the paraboloid would form the basis of the centimetre aerial system and I had found that a firm, known as the London Aluminium Company, was capable of spinning full paraboloids in aluminium. They did so, but made the mistake of sending them to me by goods train from Paddington. Day after day I was exasperated by their non-arrival. Somewhere on that 100 mile journey they had been lost. Through these days of July there were repeated air battles near us as the Germans bombed Southampton and Portland and, by defying the orders to descend into the air raid shelters on our field site, we witnessed the initial stages of the great air battles of that summer and were strangely elated as the Heinkels spiralled in flames into the blue sea. However, their proximity to the site of AMRE over which they often flew and machine-gunned was soon to end our stay in that idyllic place.

Figure 4.4 The E1189 cavity magnetron manufactured by GEC and first used at Worth Matravers July 1940. (Reproduced by kind permission from E G Bowen *Radar Days* (Bristol: Adam Hilger).)

† For further details of this equipment and a photograph (the most obvious feature is the twin paraboloids) see Burcham and Shearman.[14]

Before that happened the first of the sealed-off Randall–Boot cavity magnetrons arrived from GEC on 19 July and, a day later, my spun paraboloids at last appeared at Worth having taken several weeks on the journey from London. The paraboloids had an aperture of three feet and on 22 July I measured the polar diagrams and found that by moving the dipole across the aperture I could get a beam shift of ± 25 degrees without too serious a deterioration in the beam shape. The problem of using a single aerial for transmitting and receiving still awaited a solution and I began experimenting with a split version of the full paraboloid.

Throughout these days we had many official visitors culminating on 30 July (a Tuesday) with the C-in-C Fighter Command, Sir Hugh Dowding. He looked at our 10 centimetre devices and then assembled us to give a talk on AI, the gist of which was that the most important thing in the world was the correct adjustment of the 1½ metre azimuth aerials on the Blenheims which he said he must have by the end of the week. It would be hard to imagine a more inappropriate talk since we had no connection with the azimuth aerials on the Blenheims. However the depression caused by his visit soon gave way to elation.

As soon as the sealed off cavity magnetron arrived on 19 July, plans were made to wire it into a crude pulsed system as quickly as possible. Atkinson and Burcham had been working on the pulsing of the klystron and now had the job of making the modulator to pulse the magnetron. Skinner collaborated with A G Ward in making a receiver using a crystal mixer and on 8 and 9 August, with the equipment connected to fixed paraboloids, echoes were obtained from the coastguard hut.† In the absence of any common transmit–receive device I asked the workshops to make a swivel on which to mount the two three-foot paraboloids. By this time Rowe had placed Dee in overall charge of us and with only one piece of equipment a good deal of confusion reigned. However, on 12 August, the day I got the double paraboloid swivel from the workshop, Dee, Skinner and Atkinson were away and the site was unusually quiet, Burcham made cables to carry the magnetron power to one paraboloid and the receiver to the other. At 6 pm we were ready. By chance an aeroplane was flying along the coast a few miles away and from that unidentified aeroplane we received the first 10 cm echoes using the cavity magnetron. The next day (13 August) there were more echoes from aircraft. Dee had returned and brought down Watson-Watt and Rowe to see these historic echoes. In the afternoon we sent one of our junior assistants, Reg Batt, with a tin sheet and told him to cycle along the cliff in front of us. The ground rose slightly to the face of the cliff and where that young

---

† Although my diary refers to the coastguard hut this could not have been the case because it was below the ridge of the cliffs as seen from our site. The echoes must have been from the coastguard houses and St Aldhelm's Chapel which are above the ridge.

man cycled the ground was behind the tin sheet as viewed from our paraboloids. As we swivelled the paraboloids to follow Batt and the tin sheet a
strong echo appeared on the cathode-ray tube. I merely noted that it was
'amazing considering it should be right in the ground returns'. None of
us had any inkling on that afternoon of the immense significance of that
somewhat casual experiment.

The Last Days at Worth Matravers

As the great air battles of that August developed over the south of England,
the work at Worth Matravers began to suffer serious disruption. We were
an obvious target because an operational CH station had been built across
the lane from the main AMRE site. On the cliffs just two fields from our own
field site the experimental CHL station was in operation. One day I was
watching the CRT with Rowe when, without warning, we heard the
alarming swish of a bomb which fortunately missed the CHL and fell into
the sea. On the twenty-third, when I was measuring polar diagrams, a
Heinkel III flew over the site at 1000 feet. We were an easy target but
German intelligence had completely failed to penetrate the veil of secrecy
surrounding our work. Even so, there were often two air raid warnings
during the day causing us, under orders, to dash for the shelters and the disruption to the work became serious.

Speculation and rumour arose that, once more, the establishment was to
be evacuated to another place. In the event this move was not a major one.
On 24 August a notice was circulated that we were to move to a
school—Leeson House—in Langton Matravers, about half-way between
Worth and Swanage where most of us were living. Immediately, with
Atkinson, I went to Leeson and we decided that a stable block of the school
and the associated pigsties would be an ideal site for us. It was on the edge
of a field with a view over the town of Swanage and across Swanage Bay to
the Isle of Wight. The choice was soon to prove better than we realised on
that August day. After much manoeuvreing among the senior people, in
mid-July Rowe had formally placed Dee in charge of this centimetre group.
He was ill with a high temperature but we penetrated to his bedside and persuaded him that the Leeson stable site was the only possibility if we had to
move from Worth. Within a few days the decision was made but immediately the tension about an invasion appeared likely to eject us away from
the coast. In the evening of 11 September Churchill warned the nation to
expect an invasion. As Tizard had correctly foreseen in 1936 the Luftwaffe,
having been defeated by day, turned to night bombing and the great blitz
on London had begun, while the RAF constantly bombed the invasion
ports.

On Sunday 22 September we received the invasion alert. We strapped a mattress on the car-roof as a protection against machine-gun bullets and filled the car with essential items for survival. When we could force no more into the case I sowed a row of carrots—such was the innate optimism in the face of the long odds of those days. We did the same at Worth. The essentials of AMRE were ready to be taken away in a convoy. On our field site we did not rate our chances too highly and decided to pack our most secret equipment and files into two trailers and drive them to Leeson House where we thought there would be a better chance of joining the convoy. With that completed I continued attempts to measure the side lobes of the paraboloids with equipment that would be of little interest if captured by the Germans. The evening came, fierce gales continued to rage in the channel so we went to the cinema and 'began to feel that the imminent invasion warning was a bit of a racket'.

For the rest of September we worked both at Worth and Leeson, improving the feed system to the paraboloids, getting the impedances matched and generally making a substantial improvement in the performance of the system. By the time we left our field site at Worth completely in the first week of October we were easily getting ranges of six miles on a Blenheim.

References

1 Burcham/Lovell correspondence December 1990/January 1991 deposited with the Lovell papers in the John Rylands University Library of Manchester.
2 Hansen W W 1938 *J. Appl. Phys.* **9** 654
3 Varian R H and Varian S F 1939 *A high frequency amplifier and oscillator* in *J. Appl. Phys.* **10** 140, 321–327
4 See Guerlac H E 1987 *Radar in World War II* (Tomash Publishers and the American Institute of Physics) vol 8 *The history of modern physics 1800–1950* p 196
5 A brief history of these developments can be found in reference 4 p 187 *et seq*; in Swords S S 1986 *Technical history of the beginnings of radar* (London: Peter Peregrinus) on behalf of the Institution of Electrical Engineers 1986 p 258 *et seq*; and in Callick E B 1990 *Metres to Microwaves* (London: Peter Peregrinus)
6 Randall J T 1946 *The cavity magnetron* in *Proc. Phys. Soc. (Lond)* **58** 247. Further details of the evolution of the first cavity magnetrons and their characteristics have been given by Boot H A A and Randall J T 1946 *The cavity magnetron J. IEE* **93** pt IIIA p 928. A description of the events leading to the discovery of the cavity magnetron has been given in a booklet prepared for the symposium to celebrate the 50th anniversary of the discovery. Burcham W E and Shearman E D R 1990 *Fifty Years of the Cavity Magnetron* (University of Birmingham)
7 See Callick E B 1990 *Metres to Microwaves* (London: Peter Peregrinus) p 66
8 For details of the CVD and GEC involvement in the manufacture of these first cavity magnetrons see Callick E B 1990 *Metres to Microwaves* (London: Peter Peregrinus)

9   Wilkins M H F 1987 *John Turton Randall* in *Biog. Mem. R. Soc.* **33** 493 section
    on *The Cavity Magnetron* pp 502–510

10 Callick E B 1990 *Metres to Microwaves* (London: Peter Peregrinus) p 62 *et seq*
    and Burcham W E and Shearman E D R 1990 *Fifty Years of the Cavity Magnetron*
    (University of Birmingham) pp 16–17

11 Callick E B 1990 *Metres to Microwaves* (London: Peter Peregrinus) p 62

12 Dee P I *A personal history of the development of cm. techniques* AI, ASV, *H_2S and*
    AGLT (a copy is in Lovell TRE record)

13 See reference 1

14 Burcham W E and Shearman E D R 1990 *Fifty Years of the Cavity Magnetron*
    (University of Birmingham) pp 12–13

Chapter 5

Leeson House

The first diversions from total concentration on producing a 10 centimetre AI system for night fighters began during the early autumn at Leeson House. On 22 September, before our main equipment had been moved to Leeson, Air Chief Marshal Sir Philip Joubert de la Ferté, then Assistant Chief of Air Staff, came for a demonstration and made the discouraging comment that the 1½ metre AI seemed satisfactory and we should give priority to devising a 10 centimetre Gun-Laying (GL) equipment.† Quite soon this became involved with internal politics as the senior members strove for superior authority. Dee, in particular, had been in constant discussion with Rowe and Lewis about the job he was expected to do and on 8 August, when Sir Frank Smith, then the Controller of Telecommunications Equipment, Ministry of Aircraft Production, visited Worth he had complained bitterly about the lack of a coordinator. The trouble was not only internal. GEC had the official contract for a short wave AI system and, having concentrated their efforts on a 25 centimetre system using the micropups, were obstructing the change to 10 centimetres. At Worth on 22 August we had demonstrated the 10 centimetre magnetron system to D C Espley and G C Marris‡ from GEC and my diary records that they were:

> ... very much sobered down by Hodgkin's presentation apparatus (see below). Then in the afternoon we managed to follow a Battle for 2 miles tail on which was magnificent.

† J R Atkinson has made the comment (private communication) that Lewis persuaded Joubert to return the next day in mufti and after talking to Dee and Skinner he supported the idea of 10 centimetre AI.

‡ I am indebted to Sir Rober Clayton for the following information about Espley and Marris. Both were members of the Leading Scientific Staff of the GEC Research Laboratories at Wembley (there were only two staff grades for graduates at that time in the Laboratories—Scientific Staff and Leading Scientific Staff). Marris was responsible to Paterson, the Director, for substantially all work in the Laboratories on communications and broadcasting. Espley was responsible to Marris for work on television. Soon after the outbreak of war the team on television at Wembley was at the Government's request diverted to work on airborne radar and Espley led this team, continuing to report to Marris.

Dee's complaint to Sir Frank Smith had the result that, before the end of August, Rowe had been instructed by DCD to place the centimetre AI work in the charge of Dee. A few days later Dee had exercised his authority and at a meeting with GEC representatives on 3 September, Dee recorded that:

> [GEC] did not particularly enjoy receiving our instructions. However, we made short shrift of their 25 cm proposals at this meeting and made it perfectly clear that the programme would have to be based upon a 10 cm system with a scanning parabolic mirror and a crystal mixer.[1]

The problem of scanning the narrow beam from a paraboloid on a wavelength of 10 centimetres was an issue of constant discussion. One obvious solution was to rotate the paraboloid rapidly about a vertical axis with a slower motion in the vertical plane to produce a helical scan. On 25 October at a meeting with GEC, Dee and Hodgkin agreed that GEC should develop a helical scanner which would give a beam coverage of about 45 degrees in elevation. There were two disadvantages. In the continuous rotation half the scanning time would be wasted by looking back into the aircraft, and a rotating joint would have to carry the high peak power of the magnetron. In the state of development at that time this seemed a most undesirable solution for an airborne equipment where the problem of arcing across the rotating joint would be exacerbated. GEC eventually designed a model using two paraboloids mounted back to back with a device so that only the forward-looking paraboloid transmitted and received power. Their estimates of the development time far exceeded that of the target dates then being set for testing a centimetre airborne system 'by Christmas 1940' and more faith was placed in proposals then being developed with the firm of Nash & Thompson at Surbiton.

This association with Nash & Thompson, which was to have far-reaching consequences for airborne centimetric radar, arose by chance in mid-July. This was a month before the first centimetre echoes from an aircraft had been received on the crude ground system at Worth. On a day when most of the senior people were away at a conference A W Whitaker of that firm came to Worth. At Dee's suggestion Hodgkin had already thought a good deal about the problem of scanning the narrow beam and had persuaded the workshop to make a crude scanner in which the dipole vibrated vertically at about five times per second while the whole scanner oscillated over an arc of about 120 degrees every two seconds. Someone at Worth, presumably knowing about Hodgkin's scanner, thought that in the absence of the senior staff Hodgkin might be a good person to talk to Whitaker. He did so, sitting on the cliff overlooking the sea and thereby began an invaluable association. Whitaker immediately became interested in the scanning problem and offered to make an attempt to produce an oscillating scanner according to Hodgkin's specification. Hodgkin had rightly decided that rather than give the dipole a vertical motion it would be preferable to leave the dipole fixed and give both horizontal and vertical motions to the paraboloid. In November Nash & Thompson had completed a model of this oscillating scanner. Although the horizontal oscillation worked smoothly, the fast vertical motion did not. The imposition of the vertical motion on the horizontal oscillation produced a most violent vibration and noise and the idea was soon abandoned in favour of another scheme on which the firm had been working.

At the time in the summer when Hodgkin was working on the crude oscillating scanner I talked to him about the problem whilst we were driving to the aerodrome at Christchurch. Suddenly he had the inspired solution. He would keep the dipole fixed and rotate the paraboloid at high speed in a spiral motion so that the beam scanned an area in front of the aircraft. The suggestion made by Hodgkin to Whitaker was that a 28-inch-diameter paraboloid should be spun at 1000 revolutions per minute and that the eccentricity should be made to vary rapidly in order to scan the area ahead of the aircraft. Nash & Thompson made a small scale model and six weeks later produced the full scale scanner—in which the angle of tilt varied sinusoidally over ± 30 degrees. Since the dipole was fixed the beam was deflected about 50 per cent more than the axis of the paraboloid (with some de-focussing). Hence the transmitted beam swept out an area of about ± 45 degrees. This was the scanner which Hodgkin first flight tested in March 1941 (see p 63–4). For the spiral scanner eventually used in the first operational centimetre AI system (see Chapter 6) Hodgkin[2] specified a tilt varying between 1.5 and 31.5 degrees and, since the dipole was fixed, the total cover ahead of the aircraft was about ± 47 degrees. This spiral scanner was first flown in June and July 1941 (see p 65), and as will be related, solved the scanning problem for the first of the operational centimetre AI systems. Hodgkin had also visualised how to display the target information to the pilot. During that drive to Christchurch he thought it would be possible to couple a magslip to the spiral scanner and drive the time base of a cathode-ray tube in synchronism with the spiral motion. He thought the target aircraft would appear as an arc of a circle with a radius corresponding to the range and in a position to which the pilot had to manoeuvre. As the pilot turned towards the target Hodgkin thought that the arc of the echo would lengthen and eventually became a complete circle when the target was dead ahead. As the pilot closed with the target, so the radius of the circle would decrease. When the system became operational this presentation was found to be readily readable by the RAF operators.

The Gun-Laying (GL) Diversion

Dee quickly discovered that the severe engineering and technical difficulties of producing an operational AIS system (as this was now known) was a small part of his overall problem. He had been absent since 11 September with an attack of pneumonia and when he returned to Leeson House on 21 October he found that:

> AIS had gone into almost total eclipse. Apparently someone has sold the idea that GL can easily be done on cm.

The metre wave GL system for the army anti-aircraft guns had been the responsibility of another establishment (ADEE†—the Air Defence Experimental Establishment) and Dee discovered that Rowe was 'seizing this opportunity to try and filch the GL problem from ADEE' and that:

> ... only Hodgkin is carrying on undisturbed with AIS, and Lovell and Ward are fortunately engaged upon basic work with aerials and receivers and are therefore relatively undisturbed by this new flap.[3]

Dee's opinion that Rowe was seizing the opportunity to develop the centimetre GL system in his establishment may have had substance. The development of the klystron in Birmingham by Oliphant's team was undertaken with the GL problem primarily in mind. In this development Oliphant's group was linked with Messrs BTH in Rugby and on 8 October (1940) Oliphant pressed the Ministry of Supply to order 12 copies of the prototype centimetre GL (using the klystron) as it then existed. A contract was placed but only one equipment was ordered.

I do not know to what extent Dee was aware of this Birmingham/BTH development, but, concerned about our diversion from centimetre AI, he protested to Rowe who ruled that centimetre GL must have priority over AIS for the next one to two months. After the visit of Joubert on 22 September Rowe had placed D M Robinson‡ in charge of producing a centimetre wave GL system. Robinson's task was a formidable one. In order to assemble such a GL system he had to extract from the various specialists, who were developing transmitters, receivers and aerials, the essential units to make a complete system so that the diversion to the GL programme was even greater than Dee realised. It is true that I was engaged on basic work with aerials, but it was work directed to the GL problem—that of producing a split-beam. Already at that time I had a dipole offset from the focus and rotating at 6000 RPM. The magnetron power was taken through a rotating 'capacity sleeve' and the main difficulty was to get the high-pulsed power to

---

† In 1942 ADEE became RRDE—the Radar Research and Development Establishment, and eventually ADRDE—the Air Defence Research and Development Establishment.

‡ Robinson had joined TRE in 1939 as an engineer of considerable experience. For the previous eight years he had been a research engineer with Callendars Cable & Construction Co., and since 1935 he had been at the Scophony Television Laboratory.

the rotating dipole without too severe a power loss through this sleeve. On the day of Dee's return Rowe had brought an American technical mission and Sir Ludlow Hewitt, Inspector General of the Royal Air Force, to see our magnetron equipment. Afterwards we were told by Rowe that the centimetre GL had to be ready so that an ack-ack gun at Wareham could be fired with its use in a fortnight. Although indicative of the sudden enthusiasm for centimetre GL the idea was quite unrealistic. By 6 November Robinson had managed to assemble a complete set of units for the centimetre GL at Leeson but, on 25 November, he minuted[4] Rowe and Lewis demanding 'that the promise [to fire a gun assisted by centimetre GL] be acknowledged as premature, in view of the present state of the 10 cm receiver technique.' In the previous 19 days he listed 17 days lost—five caused by high voltage breakdown because of rain blowing into the leaky trailer, six because of receiver failure, four because of the slowing down of effort caused by persistent rain and slippery mud and two because of problems with the rotating aerial joint. The troubles with the capacity sleeve were soon overcome and the rotating dipole system gave good ranges on aircraft, but the whole device was experimental and unsuitable for use by the army. In December D M Robinson was ordered to take the device to BTH in Rugby so that an engineered version could be produced.

The euphoria engendered by the success of the first experimental 10 centimetre systems using the cavity magnetron obscured the gulf that separated this research equipment from the possibility of operational use. On 30 December Dee made this comment in his diary record:[5]

> The GL fiasco has ended up with the whole thing being moved *en bloc* to BTH, including two AMRE staff. Nothing ever worked properly at Leeson and Robinson is feeling that it has been very salutory for Lewis to learn how haywire all the basic technique really is.

One result of this move was that the specification in the Ministry of Supply contract to BTH was changed in January 1941, from Klystron to Magnetron and eventually on 31 May (1941) the BTH version of GL arrived at ADRDE (Air Defence Research and Development Establishment). According to D H Tomlin[6] who was a member of the centimetre GL development group at ADRDE, a specification was then drawn up for the main production of this system (GL3) and sent both to the USA and Canada. The British–Canadian GL3 systems eventually became operational but did not achieve the success of the parallel American development—the SCR584 which had automatic tracking facility. Many thousands of these were manufactured and were widely used by the British Army.†

† See E G Bowen[7] and the references in H E Guerlac.[8] SCR584 was first used operationally by an AA Battery protecting London early in 1944. 39 sets were landed on the Normandy beaches on D-day. The AA guns also used the SCR584 with significant success in the defence against the V-1 flying bombs.

Apart from devising the rotating dipole system to give a split-beam from the paraboloid I had not been involved in this diversion in TRE from the air interception problem, and, in fact, my rotating dipole device soon became part of another version of the centimetric AI. However, before that happened I was involved in another and far more successful diversion.

The Naval Diversion

The early autumn months at Leeson coincided with slowly-increasing numbers of magnetrons and associated equipment for the work on 10 centimetres and with many recruits to Rowe's establishment. Quite soon two other schools were requisitioned. A boys school, Durnford House, also in Langton Matravers and Forres in Swanage. The latter served as a radar school. Rowe moved his office and the main administrative staff to Durnford together with part of the research group.

The work on centimetres remained at Leeson House and soon we had three trailers of 10 centimetre systems using the cavity magnetron working in the field, in front of the stables. One of these was used mainly by Hodgkin with the embryonic equipment destined for flight trials of centimetre AI 'by Christmas' and initially using his crude oscillating scanner. Another housed the experimental centimetre GL equipment destined to be taken to BTH Rugby in December by Robinson. In the third, central trailer, nominally in my charge, we were attempting to maximise the performance. In fact, the development was so rapid that individuals soon became authorities on particular aspects of the equipment—for example, Skinner and A G Ward on receivers and mixers, Atkinson and S C Curran (who had joined the group at Leeson) on modulators for pulsing the magnetron. Burcham was the expert on magnetrons and cables and also had the troublesome responsibility of providing the 80 volt 1000 cycle supply (on which the airborne equipment worked).

The first of the centimetre groupings emerged during the summer months at Worth. Skinner was the senior in the group dealing with the fundamental aspects of the system. In particular he had A G Ward concentrating on the receivers, the mixers and local oscillators, and A T Starr on the vital problem of evolving a system so that the same paraboloid could be used for transmitting and receiving. Dee had been firmly instructed by Rowe and Lewis to be responsible for fitting a Blenheim night fighter with a centimetre AI equipment. On this problem at that time he had Hodgkin, Burcham and myself. However, although I was to remain closely associated with Dee for the remainder of the war it was, as will be related, mainly with Hodgkin, Burcham and GEC that Dee accomplished the formidable task of making the crude ground equipment into a truly airborne interception system.

My own separation from the immediate problems of AIS began on October

28 when two naval commanders came to see the equipment in the trailers. The trailer for which I was primarily responsible was the effective test bed for extracting the maximum performance from the centimetre equipment. I had the two 36 inch paraboloids on a swivel with which Burcham and I had detected the first centimetre aircraft echoes at Worth in August. The transmitter was a GEC cavity magnetron E1189† working on 9.1 centimetres with a nominal pulse output of 5 kilowatts. The receiver used a crystal mixer and the standard 45 megahertz intermediate frequency unit developed for the 1½ metre AI. Now, in front of the stables at Leeson, 250 feet above sea level, we had a view over the town of Swanage and across Swanage Bay to the Isle of Wight. We were achieving ranges of 12 miles on aircraft and could easily see the echoes from ships in the Bay and from the cliffs of the Isle of Wight about 20 miles distant. My own brief diary entry referring to 28 October‡ 1940 simply states:

'[Monday] visit of two naval commanders who opened their arms at 10 cm for submarine detection. Of course we are extremely keen on it'.

The 'we' in this note probably refers to Skinner, who had been in charge of the Sumburgh Head radar installation before AMRE moved to Worth Matravers. In fact it may well have been Skinner who arranged the visit of the two naval personnel. It was certainly not Dee for on that day his diary entry reads:

From Leeson House we recently demonstrated 10 cm equipment viewing ships in Swanage Bay. Horton and Willett were at first very defeatist about this saying that from the Naval point of view the war would have to be won on 3.5 m. Skinner and party however are strenuously advocating fitting a 10 cm equipment to a destroyer for anti-submarine work. This is yet another interruption to the AI programme since very little work is now going on on the basic technique of cm ... surely it would be a lot wiser to get a reliable Tx [transmitter] and Rx [receiver] going first ...

Captain B R Willett RN was, at that time, the Experimental Captain in charge of the Experimental Department of HM Signal School§ and C E Horton was head of the RDF section. Although it is uncertain both from Dee's diary entry quoted above, and my own reference to 'two naval commanders' whether it was to Willett and Horton that the demonstration was given on 28 October, I am advised by J R Atkinson[9] that this was the case and that their visit had been arranged by Skinner.

† See p 58 for the difference between the magnetrons E1198 and E1189.
‡ According to Burcham's diary the visit was on Sunday, 27 October 1940.
§ Later (26 August 1941) the Admiralty Signals Establishment (ASE).

In any event after the 28 October demonstration affairs moved quickly. On
8 November†

> we had a naval contingent to see a very impressive demonstration on boats
> and as a result of a conference they are sending a few men to help us with
> a prototype. So we are very pleased ...
> [The boat was the Titlark of about 92 tons].

Dee's diary states that on 7 November we had a:

> visit of several Naval big bugs to discuss the submarine problem. A
> submarine was demonstrated at 9 miles and we managed to enthuse them
> quite a lot.

In this diary entry Dee has assimilated two different events—the demonstration on the boats on 8 November and that against the submarine on
11 November‡ when

> we gave a superb submarine demonstration which seemed to shake Willett
> and Horton pretty much.

The submarine was *HMS Usk* and in Skinner's report[10] it is recorded that
on 11 November a range of seven miles was observed on the surfaced
submarine and four and a half miles on the conning tower. Of this event
Alan Hodgkin recalls[2]

> Both Willett and Horton thought it right to make an extremely conservative
> estimate of the range at which an echo from the submarine could be seen
> against noise. Radar echoes fluctuate as the aspect of the target changes, so
> there was room for disagreement, which is what we had. As the submarine
> approached Swanage, Lovell insisted that the echo from it was plainly visible
> when it was quite a long way away but Horton and Willett refused to make
> such an entry in their notebooks until the echo was so large that even
> the most casual observer could detect it. I don't think I actually witnessed
> this scene, but Lovell's subsequent indignation is clearly engraved on my
> memory.

In the event, Willett and Horton were sufficiently impressed to send over
a group of scientists from ASE to help us build a prototype apparatus in a
trailer for their further tests. By 8 December we were able to track the

† According to Skinner[10] this demonstration was on 7 November. My own diary note
written on Sunday 10 November refers to 'Friday morning' as the time of the demonstration and this places the date as 8 November. Burcham's diary agrees with the
8 November date.
‡ The date of the demonstration against the submarine (11 November 1940) is
confirmed in Skinner's report.[10]

Titlark launch to a range of 13 miles. We then towed the trailer to Peveril Point above the town of Swanage and then to the pier at Swanage so that we could test the performance at heights of 50 feet and 20 feet above sea level. These tests had been made using the two paraboloids on a swivel which produced a narrow 'pencil' beam, but the ASE scientists pointed out that this arrangement would be useless for a shipborne installation since every time the ship rolled the target echo would be lost. For the last test (on Swanage Pier) we replaced the two paraboloids by the 6 foot × 9 inch aperture cylindrical (sectional) parabolas which I had used at Worth. This retained the narrow azimuth beam but produced a fan beam of about ±15 degrees in elevation. The results were excellent and these sectional parabola or 'cheese' type aerials were to become a standard form of aerial for use in smaller ships.

> (Dec. 24,1940) It was very sad when they removed it [on 19 December 1940] to Portsmouth ... we felt, in spite of our own battles, that the work was now going to sink and hang fire.

This pessimism turned out to be quite unjustified. In March 1941, only three months after the trials on Peveril Point with the experimental prototype, fully engineered sets were on sea trials in the corvette *HMS Orchis*, and late in August a system was operational in the battleship *HMS Prince of Wales*. On 16 November 1941, only one year after the demonstration of the equipment at Leeson House, the sinking of the U-boat U433 near Gibraltar was directly attributable to its detection by the 10 centimetre radar in the corvette *Marigold*.[11] By May 1942, 14 months after the sea trials in *HMS Orchis*, 236 ships fitted with the 10 centimetre radar were at sea.

Neither in my diary notes nor in those of Dee are the scientists from ASE who came to help with the installation in the trailer identified. However, in their papers on the development of naval radar Coales and Rawlinson[12] identify them as S E A Landale, C A Cochrane , J R Croney and C S Owen. This equipment was known as the Admiralty Type 271 and by the end of the year 50 were operating in naval escorts.

The Type 271 equipment and the subsequent development of improved versions by the Admiralty Signals Establishment became of great importance to naval operations and the account given here concerns only my association with the experimental prototype equipment during the period late October to December 1940. Further historical research on that important period carried out by Commander Derek Howse during the writing of his *New Eyes for the Navy*[11] is in close accord with this summary.

Those autumn months of 1940 were in sharp contrast to the frustrating period at St Athan a year earlier. Now we were resident in a pleasant town and some, at least, of the pre-war social life returned as increasing numbers of scientists of diverse interests joined Rowe's establishment. However, on almost every night, the German bombers crossed the coast on their way to their devastating attacks on Bristol, Coventry and other cities in the Midlands. Awoken night after night by the sirens one listened for the swish of the falling bombs. The great success of the RAF in combating the daytime attacks of that summer were not repeated and throughout those months the Luftwaffe continued its attacks at night relatively unmolested.

On 22 December Sir Philip Joubert de la Ferté, during one of his many visits, protested vigorously against the lack of effort on AIS as compared with the Admiralty and GL problems. This was in sharp contrast to his remarks precisely three months earlier that we should concentrate on GL and it is to the AI programme and an associated development in which I became involved that I must now turn.

References

1 Dee P I *A personal history of the development of centimetre techniques AI, ASV, H₂S and AGLT* (in Lovell TRE Record)
2 Hodgkin A L *Chance and Design: Reminiscences of Science in Peace and War* (Cambridge: Cambridge University Press) to be published
3 See reference 1
4 Copy of minute DMR/NC 25 November 1940 in personal communication from J R Atkinson deposited with the Lovell papers in the John Rylands University Library of Manchester.
5 See reference 1
6 Tomlin D H February 1990 Lecture to the Symposium at the University of Birmingham *Fifty years of the cavity magnetron*
7 Bowen E G 1987 *Radar Days* (Bristol: Adam Hilger) pp 191–192
8 Guerlac Henry E 1987 *Radar in World War II* (Tomash Publishers and the American Institute of Physics)
9 Atkinson J R Personal communication 10 January 1991 deposited with the Lovell papers in the John Rylands University Library of Manchester.
10 Skinner H W B *Preliminary report on tests on ships using 10 cm waves November–December 1940*. TRE file 4/4/457 20 December 1940
11 Howse H D 1991 Personal communication
12 Coales J F and Rawlinson J D S 1987 *The development of Naval Radar 1935–1945* pt 1 in *J. Naval Science* **13** No 2 p 125

Chapter 6

AIS—the First Centimetre AI (AI Mark VII/Mark VIII)

When Joubert came to TRE† on 22 December 1940 to make his vigorous protest about the lack of progress on AIS, the efforts of the Army anti-aircraft guns and the RAF night fighters to protect the country against the night bombing were meeting with little success. London and many provincial towns and cities were under constant and heavy attacks at night. The losses of the German night bomber squadrons were very small—less than 1 per cent of sorties. Furthermore, of those night bombers that were destroyed the anti-aircraft guns had the monopoly of kills.

The Operational Use of 1½ Metre AI

The 1½ metre AI in the Blenheim night fighters had little success. However, in the light of future events it is evident that the problems did not wholly lie in the limitations of that particular radar equipment. The Blenheim night fighters could not compete with the performance of the German night bombers, but towards the end of 1940 the more powerful Beaufighter aircraft began to enter service use. At that period, also, the difficulty of locating the target aircraft by the 1½ metre AI because of the maximum range limitation (height above ground) was lessened with the advent of the Ground Controlled Interception (GCI) system. The work of many people since the early months of the war had led to the development of the plan position indicator (PPI) and this display technique had been demonstrated during the summer at Worth by using it with the rotating aerial of a 1½ metre CHL radar. By the autumn the embryonic GCI system had been evolved from this combination. In this the radar echo from the target and the night fighter were both

† At about the time of the move of AMRE from Dundee to Worth in May 1940 there was a change in ministerial responsibilities and the establishment became MAPRE (Ministry of Aircraft Production Research Establishment). In November 1940 the name was again changed to TRE (Telecommunications Research Establishment).

displayed on the PPI and the ground controller could give instructions to the pilot of the night fighter to bring the aircraft within AI range of the target.

The first GCI equipment was hand made in TRE and on 18 October 1940 the first operational test was carried out with the night fighters under GCI control. A month later (the night of 19–20 November) Group Captain Cunningham, flying a Beaufighter equipped with the latest version of the 1½ metre radar using a pulse modulator (AI Mark IV) and under GCI control, engaged a German night bomber. The night after Joubert's 22 December visit to TRE Cunningham made his second successful night interception. †️ At the beginning of 1941, a dozen GCI systems were in operation with RAF night fighter squadrons and, as the training and experience of the ground and airborne operators improved, the GCI/AI night fighter combination achieved significant successes. By March 1941 this combination had changed the night battle picture. The German night bomber losses occasionally reached 10 per cent of sorties—and now it was the AI equipped Beaufighters that had the monopoly of success. ‡️ The Luftwaffe could not sustain these heavy losses and by May 1941 the first phase of the massive night bombing raids came to an end.

The First Centimetre AI

In March 1942, exactly a year after the 1½ metre AI in Beaufighters under GCI control began the RAF's ascendancy over the German night bombers, the first centimetre AI was used operationally. Using the Hodgkin/Nash & Thompson spiral scanner, and known as AIS during the development stage, the equipment as fitted in the Beaufighter night fighters was called AI Mark VII. The transference of the ungainly experimental ground equipment, using the cavity magnetron which first gave echoes from aircraft in August 1940, to the elegant operational airborne system 18 months later was a remarkable achievement. Although the major factor in making centimetre AI a feasible proposition was the Randall–Boot cavity magnetron, a multitude of difficulties had to be surmounted before it became possible to use the magnetron in an airborne system.

The development of the original experimental ground system to the first airborne installation was the result of close collaboration between Dee and Skinner's TRE staff and the GEC engineers from the Wembley research lab-

---

†️ I am indebted to Group Captain Cunningham, CBE, DSO, DFC, for this information from his log books. His AI operator was Aircraftman J Phillipson. His third successful combat occurred on 2 January 1941. In January, Squadron Leader T Rawnsley succeeded Phillipson as Cunningham's AI operator. Their first of many successful night combats under GCI control occurred on 12 January 1941.

‡️ In May 1941 the RAF night fighters destroyed 102 German bombers and severely damaged a further 172.

oratories. Once Dee had convinced GEC that their contract would have to concern the 10 centimetre system and not their 25 centimetre development, he established an excellent relationship with the firm and this was a cardinal factor in the rapid progress to a successful engineered system. Until their whole time was concentrated on the AI problem the GEC team had been largely concerned with the development of television. The group was led by D C Espley and included several people who, after the war, achieved distinction either in academic or industrial life. Amongst these George Edwards was Hodgkin's major collaborator, and the team included E C Cherry, R J Clayton and B J O'Kane. The model shop facilities at the GEC Wembley research laboratories were far superior to anything that existed in TRE and it was from this source that, in a corporate effort with the TRE scientists, a large proportion of the engineered units emerged to form the first airborne installation. These units and their subsequent modifications formed the core of all the future developments with which I was involved.

The Scanner

My own tests made at Worth in the summer of 1940 on the absorption of 10 centimetre waves in various materials showed that it would be satisfactory if the nose of the fighter could be made of perspex. In the early autumn months Dee and Hodgkin arranged with the Royal Aircraft Establishment in Farnborough to fit a Blenheim night fighter with a nose made of this material. This aircraft, Blenheim N3522, arrived at the aerodrome in Christchurch in December. Dee had directed Burcham to take charge of the installation and ground test equipment at Christchurch. He set up a ground system as a test bed for the airborne units and, with Hodgkin, J V Jelley and A C Downing, bore the brunt of equipping that Blenheim with the essential cabling and hydraulic supplies for the first experimental airborne centimetre AI.

In the previous chapter I have referred to the constant discussions that occurred about the problem of the scanning of the narrow centimetre beam and the liaison between Hodgkin and Nash & Thompson that led to the construction of the spiral scanner. Because of doubts about the viability of the spiral scanner Dee had persuaded GEC to make a helical scanner. In January 1941 two 10 centimetre systems were working in trailers in Leeson, one using the Nash & Thompson spiral and the other the GEC helical scanner. The spiral scanning system was chosen and installed in the Blenheim in February.

The spiral scanner produced by Nash & Thompson was an extraordinary device. The 28 inch paraboloid spinning at 1000 RPM with an eccentricity varying by 30 degrees in one second produced a strange effect on an observer. Of this Rowe later wrote: [1]

> It must be confessed that when R.A.F. personnel at Christchurch saw the first
> A.I. scanner system installed in an aircraft, doubts were cast on the sanity of
> the scientists. Before the system reached a speed of rotation greater than the
> eye could follow, it could be watched rotating in a curiously irregular fashion
> with the one apparent desire of escaping from the aircraft altogether. When
> however the system was found to do its job and to give less trouble than
> many devices of greater apparent respectability, the R.A.F. personnel soon
> learnt to regard it as a normal piece of equipment.

The scanner was dynamically balanced and the vibration was insignificant.
In fact scarcely a ripple could be seen in a glass of water standing on the
framework of the scanner when running at full speed.

The Magnetron Transmitter

E C S Megaw of the GEC research laboratories was responsible for developing the sealed-off version of the cavity magnetron necessary for airborne
use. Early in July 1940 the first GEC sealed-off version (E 1188) gave the
same CW output as the original Randall–Boot bench model. An improved
version using an oxide-coated cathode (the E 1189) tested a few weeks later,
operating at 8.5 kilovolts in a magnetic field of 1500 oersteds produced 12.5
kilowatts. It was the second E 1189 which we received at Worth in July 1940
and which we used in the initial tests of the 10 centimetre ground system
described in Chapter 4. The urgent demand for further samples was met by
using the chamber of a Colt revolver as a drilling rig. †

After various tests and calculations made on this second E 1189, Megaw
decided that the field strength for optimum performance was too high for the
type of permanent magnet which would be needed in an airborne equipment. He then re-calculated the design on the basis of eight cavity segments
instead of the six of the original. This variant, known as E 1198 (CV38),
which worked on a wavelength of 9.1 centimetres was the magnetron used
in the first airborne AI equipment. It was one of the initial (No 12) E 1189
magnetrons that was taken to the USA by the Tizard mission in August
1940. ‡ A full account of the development of these early magnetrons and of
the successive developments that produced magnetrons of greater power
output has been given by E B Callick. [4]

---

† These and other details of the development and production of the early cavity
magnetrons have been given by E C S Megaw. [2]
‡ E G Bowen [3] relates the confusion arising in the USA from the blueprint (6 cavity)
and the actual (8 cavity) magnetron which he took to the USA with the Tizard
mission.

The Modulator—Factors Determining Pulse Length and Pulse Repetition Frequency

With an anode voltage of 7.5 kilovolts in a field of 1000 oersteds the E 1198 magnetron would produce a peak power of 15 kilowatts into a water load with a pulse rate of 50 per second and pulse length of 30 microseconds. There are two fundamental features of an AI system which make it essential to use pulse lengths and pulse repetition frequencies (PRF) radically different from those used in the ground tests with the early E 1189 and E 1198 magnetrons.

The essential requirement for minimum range is that the pulse length must not be so great that the transmitter is still operating when the echo from the target returns to the receiver. The equivalent length of a one microsecond pulse is 980 feet and, thus, a target at a distance of half this figure would represent the minimum range for a system using one microsecond pulses. This assumes an ideal case in which the pulse rises to, and falls from, peak power abruptly and that there is no paralysis of the receiver caused by the transmitted pulse. Since a minimum range of less than 1000 feet and preferably not more than 500 feet was deemed to be necessary, so that the pilot could identify the target visually, the magnetron system had to be designed to transmit one microsecond pulses with the sharpest possible rise to maximum power and with a similarly abrupt decrease to zero power after one microsecond.

It was this problem of producing such a pulse with more than 10 kilowatts of peak power that occupied Atkinson and Burcham at Worth in the summer of 1940 and later at Leeson. A thyratron and a hard valve (pentode) were tried as a basis for the modulator. As the collaboration with the GEC engineers developed during the autumn they preferred the thyratron whereas the TRE group believed that a better pulse shape and a higher PRF could be obtained by the hard valve solution. Eventually on 30 December after considerable dispute, GEC agreed to the use of the hard valve. This modulator supplied one microsecond pulses at about 50 kilowatts peak power to the magnetron. In the initial system installed in the Blenheim, the modulator was in the rear of the aeroplane and the conveyance of this high voltage pulse to the magnetron had to be through cables, plugs and sockets. In the early airborne tests it was estimated that the peak radio-frequency output from the magnetron was about 10 kilowatts.

In November 1940 S C Curran joined the centimetre group in TRE. He had worked on pulse amplifiers before the war and his experience with Geiger counters at the Cavendish Laboratory proved of inestimable value in the subsequent development of modulators for pulsing the magnetrons. He soon demonstrated the possibility of spark gap modulators which had not hitherto found much favour with the industrial firms since it had been assumed that the pulse repetition rates of over 1000 per second would be impossible to achieve with this form of modulator. Curran brought his experience with Geiger counters to bear on the problem and was soon handling 50 kilowatts of pulse power at pulse repetition rates of over 2000 per second. It was soon realised that the Curran version of the spark gap was the best solution for modulating magnetron transmitters, at least in airborne systems. This development came too late to be included in AI Mark VII and VIII but it was used in the airborne H_2S and ASV centimetre radar systems described in this book. †

Whereas the minimum range requirement determines that the pulse length should not be greater than one microsecond, there are a number of other factors that determine the pulse repetition frequency (PRF). The received echoes are displayed on an intensity-modulated (CRT) screen and, in principle, it is desirable to build up as many returned echo pulses as possible since the signal to noise ratio will improve as the square root of the number of echo pulses integrated. However, the narrow beam is being scanned at high speed and so successive echo pulses are spread out across the screen proportional to the scanning speed. At the least there must be overlap on the screen between successive pulses and hence there is a requirement that the PRF should increase proportionately with the scanning speed. As the transmitted–received beam-width decreases, so the PRF must increase so that a number of echo pulses may be received as the beam sweeps over the target.

All these factors suggest that the PRF should be as high as possible, but there are two factors which set a maximum limit to the PRF. One is the permissible maximum mean power dissipated in the transmitter, and the other is the confusing effect of echoes from a target at long range appearing on the next sweep of the time base at a short range. For example, if the PRF is 2000 per second the range of the time base is 50 miles. If a target is 45 miles distant the echo will appear near the end of the time base. However, if it is 55 miles distant the echo will appear near the beginning of the second sweep of the time base at an apparent range of five miles. Taking account of these various factors the PRF of the first centimetre AI radars was chosen to be 2500 per second.

In the AI system, with the spiral scanner spinning at 1000 RPM, the beam swept out a spiral of semi-angle of about 45 degrees (see p 47). With the PRF of 2500 per second there were ten scanning lines on the PPI on the outward spiral and ten on the inward giving an average spacing between lines of about two degrees. With the beam-width of approximately seven degrees (to half power) there was no loss of range information between the scanning lines and there were about four pulses per beam-width at the PRF of 2500.

† The technical problems associated with the development of pulse modulators for radar systems during the war are described in a series of papers in *J. IEE* **93** pt IIIA 1043–1112 1946.

Clearly a higher PRF would have been an advantage but the limitation of mean power dissipation and second time base returns mentioned above made this impossible.

The Receiver

An intermediate frequency (IF) amplifier working on 45 megahertz suitable for airborne use had already been manufactured and formed part of the 1½ metre AI system. The problem of making a receiver for the centimetre AI system was therefore centred on the problem of the mixer and the local oscillator. Skinner made a long series of experiments with crystals during the summer and autumn of 1940. An abiding memory of the days at Worth and Leeson is of Skinner, cigarette drooping from his mouth, totally absorbed in the endless tapping of a crystal with his finger until the whisker found the sensitive spot giving the best characteristics. Ultimately a tuned crystal mixer using a tungsten point on a silicon crystal was employed in the early centimetre airborne systems. In the beginning these were enclosed in a glass envelope filled with wax. The experiments were refined by Oliphant in Birmingham, and in the flight trials the crystal was mounted in a capsule.

Although only small amounts of power were needed for the local oscillator it had to operate on a frequency in the 10 centimetre band only 45 megahertz different from the frequency of the magnetron. After various trials with a split-anode magnetron and the klystron an appeal was made to R W Sutton, an authority on the design of thermionic valves and then working with the ASE.† By October 1940 he produced a version of the klystron, in which a single resonator is used for both bunching and catching. In this tube known as the reflex klystron the electron beam is reversed, after the first transit of the cavity, by reflection at an electrode held at a negative potential. By suitable choice of the anode potential the correct phase can be established so that each electron bunch arrives back after reflection in the correct phase to give up energy at the cavity. An important feature of Sutton's version was that part of the cavity resonator was external to the vacuum and this greatly facilitated tuning and coupling to the load.‡ A power output of 300 milliwatts was readily attained in the centimetre waveband and many thousands of these reflex klystrons were built into operational centimetre radar systems as well as into the original airborne centimetre AI.

---

† R W Sutton joined the valve division of the Admiralty Signal School in Portsmouth early in 1939. In 1940 the valve division was evacuated to the H H Wills Physics Laboratory, University of Bristol, and it was from there that Signal School reported that they had obtained 200 milliwatts from a reflector klystron.

‡ A detailed description of the evolution and properties of this type of klystron is given by A F Pearce and B J Mayo[5] and by E B Callick[6].

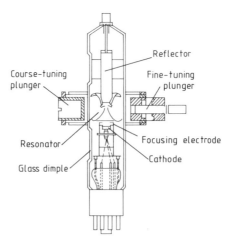

Figure 6.1 Diagram of the reflex klystron originally produced by R W Sutton as a local oscillator for centimetre AI. (After Pearce and Mayo 1946 *J. IEE.*)

The Common Transmit–Receive (TR) System

Neither the GL nor the ship applications placed such severe restrictions on space as the nose of the night fighter. In the former cases separate aerials could be used for transmitting and receiving. It was obvious during the early spring days at Worth that there would be room for only a single scanning paraboloid in a night fighter and this would have to serve both as the transmitting and receiving aerial. The problem of protecting the receiver from damage during the transmission of the high power pulse from the magnetron proved difficult to solve. Many solutions were tried and discarded. At Worth I had tried a method using one paraboloid with the transmitting and receiving dipoles separated by a metal sheet. This gave considerable protection to the crystal mixer. In the form of a circular disc of about five inches diameter this was the method used in the ground tests at Leeson throughout the autumn. However it was not satisfactory. On 30 December 1940 Dee recorded:[7]

> We are still having great trouble with crystal protection ... The crystals seem to burn out after a few hours running although some go down immediately ... This disc method of aerial protection is probably unsound and Espley of the GEC is advocating a hybrid device which might be better.

This method of protection was, indeed, unsound because, apart from the burning out of the crystals, the disc had a deleterious effect on the polar diagram of the paraboloid. In February 1941 a decision was made in favour of the hybrid solution and the first airborne installation in the Blenheim used a hybrid ring of quarter wave coaxial lines and dummy loads. Three quarters of the power was wasted in this arrangement and the lack of protection afforded to the crystal mixer was a constant source of trouble.

The solution to this problem was eventually found by A H Cooke, a member of the group at the Clarendon Laboratory, Oxford, led by J H E Griffiths. He suggested that a reflex klystron should be used without the electron gun and reflector electrode, with the tube filled with gas at a suitable pressure. The original experiments were made with the tube filled either with helium or hydrogen at a pressure of a few millimetres. This 'soft Sutton tube' or rhumbatron switch as ultimately produced was the CV43 filled with water vapour at six millimetres pressure, which flashed over during the transmitted pulse, but not when the weak echo was received. As in the reflector klystron the outside of the cavity resonator was free for coupling and tuning. The first of these rhumbatron switches was built into a coaxial line circuit by Skinner, Ward and Starr so that the gas discharge resulted in the receiver input line presenting a high impedance to the transmitter during the pulse. This provided excellent protection to the crystal from the high-powered pulse transmitted by the magnetron. The device was first used in the experimental centimetre AI in the summer of 1941 and proved so successful that, built either into a coaxial line or waveguide arrangement, this rhumbatron switch became a standard feature of the centimetre radars where the same aerial had to be used both for transmitting and receiving. A description of this rhumbatron gas switch, the coaxial and waveguide arrangements in which it was used, and the subsequent development for shorter wavelengths and higher pulse powers has been given by Cooke, Fertel and Harris[8] and by E B Callick.[9]

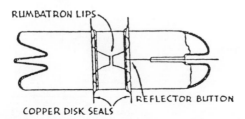

RUMBATRON LIPS

REFLECTOR BUTTON

COPPER DISK SEALS

Figure 6.2 The 'soft Sutton tube'—the key element in the common transmit–receive system for centimetre radar. (After figure 8.18 Guerlac H E 1987 *Radar in World War II* (Tomash Publishers and American Institute of Physics.)

The Flight Trials of AIS

Early in March 1941, the first centimetre AI tests took place in Blenheim N3522. Hodgkin and Edwards where the operators of the AI and after some troubles with fuses they made the first airborne centimetre AI contact with a target at a range of between one and two miles with the Blenheim flying less than a mile above ground. Soon ranges of two to three miles were achieved on a Battle aircraft as target at a greater range than the height above ground of the Blenheim.

The display on the CRT screen was much as Hodgkin had predicted. However, there were two unexpected features of the display. One, entirely favourable, was that ground returns from the main beam showed up as an artificial horizon on the CRT which rose and fell with the varying pitch of the spiral. This radar horizon proved of great benefit in the interception of targets. The other unexpected feature was a broad circular ring of ground returns caused by scattering. After several months' work Hodgkin and Edwards succeeded in reducing this unwanted ring to negligible proportions. The use of vertical polarisation instead of horizontal and the installation of a uniform perspex nose on the Blenheim, instead of the original which was strengthened by a wooden ring, were the main contributions to the removal of the unwanted ground-return ring.

Although these early flights proved that the main advantages claimed for the centimetre AI were justified—that is the ability to detect a target at a range greater than the height above ground, there were several difficulties. Experience with the experimental system at Leeson had led to expectations that the target aircraft could be detected at ranges up to 10 miles, whereas in these flight trials ranges of only three miles or less were achieved. The

One of the first airborne centimetre magnetron transmitter units. The magnetron is between the poles of a magnet. (Courtesy of Douglas Fisher.)

trouble was recognised to be the considerable loss of power in the dummy loads of the hybrid TR system and the rapid deterioration of the crystal mixers. Both these problems were solved in June 1941 when the receiver group produced a TR box incorporating the rhumbatron gas switch. In the early days of July 1941 this TR system was tested in flight and the range on target aircraft instantly improved to that predicted from the ground tests.

AI Marks VII and VIII

After the successful flights with the new TR system in early July 1941, events moved swiftly. In the last week of July a decision was taken to equip four Beaufighters with AIS for service trials. Twelve sets of electronic units were ordered from GEC and twelve spiral scanners from Nash & Thompson. Almost simultaneously orders were placed with GEC and Nash & Thompson for 100 sets to equip a squadron of Beaufighters for operational use. Thereby began the first of many 'crash programmes' involving centimetre radar systems.

Hodgkin had the task of persuading the Bristol Aircraft Company to produce four Beaufighters with perspex noses, and Burcham was in charge of the installation of the AI equipment in the Beaufighter (X7579) when it arrived at Christchurch on 4 September. After the flight-testing from Christchurch, in which Edwards of GEC took a prominent part, the Beaufighter was sent to the Fighter Interception Unit (FIU) at Ford on 30 November. Dee asked Burcham to go to FIU to take charge of these trials 'with strict instructions that a good report must be forthcoming from the RAF. In fact this was obtained but it involved a great deal of preliminary installation and flying from Christchurch'.[10]

At this time the problem of introducing this new type of centimetre equipment to the RAF arose in an urgent form. It was from this need that the TRE Post-Design Services (PDS) arose. Rowe[11] has described the formation and activities of this new group in TRE:

With the advent of centimetre A.I., therefore, we were faced with a major problem. We needed men whose task it was, not to originate devices but to study their behaviour under Service conditions; men capable, by virtue of a scientific training, of analysing the types of failure experienced in particular equipments under various conditions of use, of suggesting remedies and, not least, men capable of living with Service personnel at squadrons and of working and jesting with them. Above all we needed a leader for the work, one who was a first-class scientist and a first-class organizer. The combination is less rare than is commonly supposed, but the two attributes do not always go together. It was Dee who suggested Ratcliffe. Ratcliffe had been loosely attached to us at Dundee, but had later founded the radar school for the Army at Petersham. I appealed to Ratcliffe to rejoin us to undertake the task of building up a Post-Design Service. He came, and stayed until his scientific staff numbered hundreds, divided into groups for aiding the various R.A.F. Commands.

This P.D.S. work was of outstanding importance to the R.A.F. and to the scientists in the laboratories. As the war progressed the operational research sections at R.A.F. Commands inevitably and rightly extended their activities far beyond radar, and we came to rely on our P.D.S. staff for information on operations involving our equipment, rather than on the operational research staffs.

Without P.D.S. much of the work of the laboratory scientists would have been in vain.

Ratcliffe arrived in TRE on 27 July 1941 and immediately joined Burcham at Christchurch. Eventually, with this PDS group, Ratcliffe took over the responsibility for AI Mark VII at FIU. Of the immediate success of this PDS organisation Rowe[12] records that it:

> ... helped to design and construct indispensable test equipment, wrote manuals for the use of RAF personnel and actually fitted the first 36 of the centimetre AI sets in aircraft and then lived with the squadrons which used them.

Simultaneously with these trials of AI Mark VII, Dee, Hodgkin, Burcham and Curran were in constant contact with the firms who had to engineer the system in a form suitable for operational use. The specification for the Nash & Thompson spiral scanner needed little of their attention, but matters were different with the electronic units that had to incorporate the latest modifications and improvements deemed necessary after the service trials. The main manufacturer of the 100 units was GEC Coventry but a number of other firms were soon introduced—notably E K Cole, then at Malmesbury. The decision had also been taken that the centimetre AI system should be installed in the new Mosquito aircraft and this added De Havilands to the list.

The 100 sets for squadron use were designated AI Mark VII and by March 1942 nine Beaufighters were operational and the first successful combat with a German night bomber occurred in April. By December 1942 either in combat over the UK, or later in the Mediterranean theatre, the Beaufighters using AI Mark VII destroyed 100 enemy bombers.

Meanwhile in the autumn of 1941 whilst the service trials of AIS in the four Beaufighters were still in progress at FIU, the initial preparations were made for the production of 1500 centimetre AI systems. By the spring of 1942 the design of this AI Mark VIII had been substantially finalised and after constant pressure from Dee, orders were placed with E K Cole for 1000 sets and with GEC for 500. By August 1942 the first sets of this large-scale production of AI Mark VIII were being tested in TRE and the first RAF squadron was equipped in December.

This large-scale production of AI Mark VIII incorporated a number of features not available in AI Mark VII, notably facilities for interrogating friendly aircraft equipped with IFF (Interrogation of Friend or Foe), and for homing on to a centimetre beacon. The improved and more powerful strapped magnetrons gave greater frequency stability and the engineering design of the units enabled the system to be operated at 20 000 feet altitude without the problems of sparkover experienced in the earlier AI Mark VII.

The Mark VIII systems in the Beaufighter and Mosquito aircraft formed the core of the RAF AI systems by 1943. Hodgkin, Burcham and their colleagues from the GEC research laboratories in Wembley bore the major responsibility for the brilliant conversion of the ungainly experimental ground systems of 1940 to the elegant operational centimetre AIs of 1942 and 1943.

I had little association with these developments to operational use, but as will be described in the next chapter the various units of these systems became integral parts of another form of centimetre AI to which I was directed after my association with the naval personnel at the end of 1940. Hodgkin remained with the AI Mark VII and VIII developments until September 1942 when he was placed in charge of a project for the defence of Bomber Command aircraft. Burcham then assumed the leadership of the TRE group responsible for the developments of centimetre AI and he and Curran were responsible for much of the detailed contact with the firms manufacturing the various units. Dee, first as group leader and then as a superintendent, remained responsible throughout for the increasing diversity of our activities. Skinner became head of a separate group with Atkinson, Starr and Ward as senior members concentrating on the development of X-band (3 centimetre) techniques.

References

1   Rowe A P 1948 *One Story of Radar* (Cambridge: Cambridge University Press) pp 122–123
2   Megaw E C S 1946 *The high-power pulsed magnetron: a review of early developments* *J. IEE* **93** pt IIIA 977
3   Bowen E G 1987 *Radar Days* (Bristol: Adam Hilger) pp 166–167
4   Callick E B 1990 *Metres to Microwaves* (London: Peter Peregrinus) ch 4
5   Pearce A F and Mayo B J 1946 *The CV35—a velocity-modulation reflection oscillator for wavelengths of about 10 cm* *J. IEE* **93** pt IIIA 918
6   See reference 4
7   Dee P I *A personal history of the development of cm techniques AI, ASV H₂S and AGLT* (in Lovell TRE record)
8   Cooke A H, Fertel G and Harris N L 1946 *Electronic switches for single aerial working.* *J. IEE* **93** pt IIIA 1575

9   See reference 4

10 Burcham/Lovell correspondence December 1990/January 1991 (papers deposited with the Lovell papers in the John Rylands University Library of Manchester).

11 Rowe A P 1948 *One Story of Radar* (Cambridge: Cambridge University Press) p 96

12 Rowe A P 1948 *One Story of Radar* (Cambridge: Cambridge University Press) p 124

Chapter 7

Lock–Follow AI (AIF/AISF/Mark IX AI)

I do not know who first had the idea of making a centimetre airborne system so that the scanner would 'lock' on to the target aircraft. An obvious advantage was that, with the anticipated accuracy the pilot of the fighter could line up his aeroplane and attack the target by 'blind firing'. That is, given reliable means of automatic identification through the IFF beacon developments, the problem of first achieving a visual sighting would disappear. Later, another advantage was claimed for the lock–follow system and the airborne tests of this led to a tragic accident.

In essence the idea of the lock–follow was simple. By using a rotating offset dipole at the focus of the paraboloid one could produce a split-beam. By comparing the strength of the received echo from both components of the beam the paraboloid could be driven so that it pointed continuously at the target. The pilot would then align his fighter with the axis of the paraboloid and, provided he was within destructive range of his armament, could fire blind with confidence.

At the beginning of 1941 my liaison with the naval people had ended. Hodgkin and his colleagues from GEC were concentrating on making the AIS into an airborne system and it was, perhaps, natural that since I had earlier made the offset spinning dipole arrangement for the centimetre GL tests, I should be associated with the problem of making a lock–follow system. In any event I worked on AIF (or AISF as it was variously known at this research stage) for the whole of 1941.

The mention of automatic lock first appears in my diary notes on 8 March 1941—'thinking hard about automatic lock'. The thinking at that stage concerned the building of a ground version to find out if it was possible to make a paraboloid track an aircraft across the sky. I built such a system in the field in front of the stables at Leeson where the other experimental scanning systems were being tested and on the site from which we had demonstrated the detection of the ships in Swanage Bay to the naval personnel. At Worth I had procured a rotatable wooden summer house as a convenient method

69

of measuring the polar diagrams. This was transferred to the field at Leeson and by the end of March I had 'the summer house equipment going very well as a first stage to AIF'. These early tests were made by using cylindrical parabolas. These gave a broad beam in elevation so that the tracking could be tested first in one dimension—that is in azimuth by horizontal motion only. In these early tests the parabolas were moved by hand because I was waiting for the workshops to make the mounting that I hoped to steer automatically from the received echo.

I did not extract this from the workshops until 1 May, but as was to be my experience for the rest of the war the pressures were so intense that we had already engaged in discussions with two firms about the manufacture of the device. Thus on 17 April when the only parts of the experimental system tested were the rotating sleeves and dipoles, I had discussions with A Tustin the Traction Motor Engineer from Metropolitan Vickers 'who seemed to regard the locking-on hunting problem as easy'. The 'hunting' refers to a worry about making the driving system of the paraboloid so that it would respond to the input from the target echo without any oscillatory or other erratic motion. Dee was pessimistic and believed that our only hope was to move the mirror manually so, ten days after my talk with Tustin, I accompanied him to Nash & Thompson where we discussed the possibility that they would mount the paraboloid so that it could be aligned by manual control to equalise the split-beam echoes and thereby be pointed in the direction of the target aircraft.

However, at that time Dee could not have been aware of the genius of F C Williams who so readily turned his attention to the problem, nor that Tustin and Metropolitan Vickers had already applied their 'metadyne' system to the automatic control of guns. F J U Ritson had joined TRE in Dundee in November 1939 and had worked with F C Williams since that time. From the early days of the lock–follow concept we worked together to evolve a practical system. I had known Freddie Williams before the war. When I first went to Manchester as an Assistant Lecturer in 1936 he held a similar appointment in the Department of Electrotechnics—joined to the Physics Department through a dark tiled Victorian passageway. Some time before the war he had left the University and joined Rowe's small group at Bawdsey and now he was able to give the benefit of his wide knowledge of electronics to any one who needed it in TRE.

The locking of the motion of the paraboloid on to the echo from a moving target was precisely the kind of problem that interested him. Neither was the basic idea new to him. Indeed, he had solved an analogous problem in Manchester just before he left to join the group at Bawdsey. D R Hartree, then Professor of Applied Mathematics in the University, had built a mechanical differential analyser in the basement of the physics department. When Blackett arrived in Manchester as the Langworthy Professor of Physics in the autumn of 1937 he immediately became interested in Hartree's machine.

He discovered that in order to supply this machine with information, in the form of a functional relation between variables occurring in the equation, it was necessary to keep a pointer on a graph on an input table. The x-coordinate lead screw was driven by the machine and an operator rotated the y-coordinate lead screw by hand in order to keep the pointer on the curve. Blackett drew Williams' attention to the problem and they devised a photoelectric curve follower which would feed in automatically, not simply the y-coordinate of the curve, but the correct value of $\dot{y}$. A photoelectric device which followed the input curve could only do so in a succession of small steps and this was unsatisfactory for driving the integrating discs on the machine. To overcome this, they detected the slope of the input curve by using a single photo cell, illuminating two points on the curve alternately and using an AC amplifier. A mechanical device was used to give tan θ, and a variable speed gear in the form of an integrator completed the automatic system. In the light of the technical and electronic developments of that epoch, it was a remarkably successful device which followed the input curve more accurately than was possible by hand. Williams had stabilised the device by adding derivative of error to error, and this must have been a very early example of what was to become an accepted technique in wartime automatic tracking.

The TRE workshops had completed the motorised scanning-lock parabola at the beginning of May, and on 6 May my diary entry reads:

> Ritson and I spent the rest of the time trying very hard to get it working. There were a lot of troubles but by last night it was sweeping and the dipoles rotating and we were just going to lock when a bearing jammed so that set us back rather.

In the search phase the parabola would scan continuously. When an echo appeared an 'electronic gate' would be moved along the time base and maintained by manual adjustment on the echo. The scan would cease and the parabola would then lock-on and follow the target automatically. At a later stage F C Williams developed an electronic circuit so that the gate would sweep the time base continuously and automatically lock-on to an echo without manual intervention. †

> Sunday May 25—Great day yesterday and today as we auto followed for the first time! First a Blenheim and then some boats across the bay. It was all very satisfactory.

---

† The drifting or automatic strobe which would search and lock-on to the target echo had been developed by Williams and Ritson[1] for the 1½ metre AI Mark VI system in order to provide the pilot with a spot indicator of the target position. The experimental version was converted to a design suitable for production by A D Blumlein and E L C White at EMI.

In order to make the initial tests as simple as possible we had used two parabolas, (thereby avoiding the difficulties still existing at that time with the common TR arrangement). By the end of May: 'Monday and Tuesday last week were very successful days for automatic following. Then decided to put in elevation as well so dismantled the apparatus and returned to workshops'. Meanwhile, even before the first successful test I had gone to Metropolitan Vickers and Ferranti in Manchester with Williams and Hodgkin to discuss the manufacture of the equipment and again with Dee in mid-June.

At that time we were aiming for an accuracy of ±1 degree in the lock–follow mechanism which we thought would be adequate for interception. A greater accuracy of ±0.25 degree was thought to be necessary for blind firing. However, this greater accuracy would not be necessary at maximum velocities and accelerations of the aircraft since blind firing would only take place near zero relative angular velocities. Of the several possible auto-follow mechanisms we favoured the type where the output was proportional to the displacement between the target and the axis of the paraboloid. In these early ground tests we used a version of the control mechanism developed by Metropolitan Vickers. In this a split-field motor driving the paraboloid was controlled thermionically with a phase advancer in the valve cathode to alleviate the effect of the inertia of the system. Figure 7.1 is a copy of my sketch of the arrangement in an internal TRE report dated 19 June 1941.

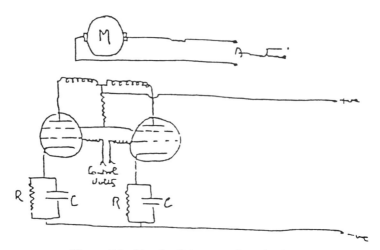

Figure 7.1 Sketch of the control mechanism.

The explanation in the report reads as follows:

The current passed by the valve with this arrangement is proportional to $V/R \div C \, dV/dt$ where V is the input misalignment volts. The field current thus leads for increasing V and lags for decreasing V. By this means periodic hunting about the zero position is very much reduced. On steady following however the system lags $\frac{1}{4}^{\circ}$ due to inertia. Thus on reversal the mis-alignment would increase to $\frac{1}{4}^{\circ}$ from a theoretical position of 'rest'. Due to the already extant forward velocity the instantaneous mis-alignment may greatly exceed this value. These figures were obtained on a steady signal. With switched and fading signals the situation further deteriorates and allowing for further possible improvements in the motor design a following accuracy of $\pm 1^{\circ}$ will probably be obtained.

Our main worry about this system was that the inertia and sluggishness of the driving motor might be too great a drawback for airborne use where we were anticipating that the lock–follow would have to work under operational velocities of 10 degrees per second and accelerations of 10 degrees per second per second. Although this basic system of Tustin's was first used in our experimental equipment at Leeson, because of the problems with the fading of the echoes and the high value of the capacitance C which would be required it was not used in the airborne equipment. The solution was devised by Williams as described below (p 75). Before Williams' solution emerged we explored an alternative system using a miniature motor which drove the paraboloid though some form of mechanical amplifier having an infinitesimally quick response—such as a torque amplifier. This was our reason for including Ferranti in our visit to the Manchester area. They had proposed to use a magnetic friction clutch. Figure 7.2 is a copy of my sketch of this proposal in the internal TRE report of 19 June 1941.

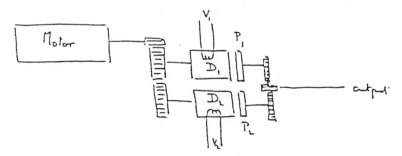

Figure 7.2 Sketch of the proposed magnetic friction clutch.

The explanation in the report read as follows:

This consists of two drums D_1 and D_2 rotated at constant speed in contrary motion. These contain small electromagnets which pull in the plates P_1 P_2 according to the direction of misalignment. This is a Class B system having a close resemblance to the regenerative torque amplifier. The chief practical

differences will probably lie in the greater time constant of the clutch
operating mechanism over the band tightening rod; and the smaller torque
available between P and D over that available from the torque amplifier
bands.

[Class B referred to an 'on–off' system in which the output was independent of the displacement of target and paraboloid axis as compared with the
Class A system of Metropolitan Vickers (Figure 7.1) in which the output is
proportional to the displacement between target and paraboloid axis.]

We left to Ferranti the problem of developing this idea and decided to give
the Metropolitan Vickers system a full-scale trial.[2]

At last the workshops finished modifying the auto-follow mechanism so
that we could test it in elevation and azimuth. On 1 July:

> our automatic follow worked marvellously on aircraft and seemed very suc-
> cessful. Unfortunately the news got to Rowe and on Sunday afternoon ... a
> Boffin party and the damned apparatus broke down.

By mid-July after another visit to Metropolitan Vickers and Ferranti,
I returned to Leeson full of hope that 50 operational Beaufighters could
be equipped with AIF by Christmas—and we had not yet even had the
equipment airborne!

There was little more we could do with the apparatus on the ground and
we began the preparations for installing the equipment in a Blenheim for
flight tests. The main transmitter and receiver units were similar to those
used in the AIS system, but in the design of the electronic circuits for the
lock–follow control two new problems arose. One concerned the fading
pattern of the received echo. Although in our early contacts with Tustin at
Metropolitan Vickers he had regarded the 'locking-on hunting problem as
easy' (see p 70) this transpired not to be the case. It would have been had
our echoes from the target been of uniform amplitude, but the amplitude of
the echoes scattered from the target aircraft were far from uniform. With the
phase advance stabilisation the servo system would not follow the target
steadily at low angular velocity. Evidently a detailed analysis of the fading
was needed. We were using a pulse rate of 2500 per second and an obvious
feature of the echo received from a target aircraft was the relatively long-
period fading and fluctuations in strength of the echo. Although this feature
was clearly visible when the echo was displayed on the cathode-ray tube we
needed data on the nature of the fading pattern for the individual pulses. On
this subject my comment in the 19 June 1941 report read as follows:

> The information for the control mechanism is obtained from a split beam
> system and the present arrangement of the apparatus gives an output of about
> 80 volts per degree misalignment of target and mirror up to $3-4°$ and then
> remains constant. This information is not, however, available in the form of

a continuous signal due to the limits set by the finite speed of rotation of the dipole. It is considered that a speed of 6000 r.p.m. can be reasonably attained which means that recurrent pieces of information can be fed to the mechanism every 1/100 sec. A further limit is set by the quick period fading of the received echo which necessitates smoothing of successive pieces of information before being passed on to the control mechanism. Sufficient information is not yet available about the rate and depth of fading to decide on the best compromise between smoothing and the rate at which information is fed to the control mechanism. Present views are that this time constant will probably be between 1/10 and 1/50 of a second, but the choice depends also on the nature and time constant of the control mechanism.

My initial attempts to obtain the data were made by photographing the echo with a cine camera running at 24 frames per second. This analysis gave the fading pattern averaged over about 100 pulses, but this information was not adequate for our purpose since the dipole was rotating at 600 RPM and we required to know the minimum number of pulses over which we could integrate in order to give the most instant control possible to the lock–follow system. In fact we required information on the pulse-to-pulse fading rate and, in the midst of my attempts to make a fast enough cine-type camera to photograph at this rate, Blackett thought that Andrew Huxley had already devised a simpler scheme. I made contact with Huxley in July and he assured me he could easily photograph data at this rate. He arrived on 17 July—apparently empty-handed and not with the complex equipment I had expected. He had a small Leica camera in his pocket and during the next few days we rapidly achieved all the information we needed. He had modified the camera mechanism so that instead of taking individual frames he could pull the film past the lens by hand. With the camera mounted in front of the cathode-ray tube, Huxley would focus on the echo and simply pull the film at a rate such that the images of the individual pulses were just separated. It took me many weeks to measure and analyse the data from the films which Huxley produced in one day by this method. In early October I had completed the reports on the analysis that indicated that a dipole rotation of about 6000 RPM would be necessary for smooth control.[3]

F C Williams devised the practical solution through which the servo would follow steadily at all angular velocities likely to be encountered operationally. He used a combination of velocity and acceleration feedback so that the driving torque per unit error was made sufficiently large at low angular velocity. In effect the acceleration feedback was used to simulate an inertia in the scanner appropriate to the angular velocity so that steady motion was achieved. Williams achieved this brilliant solution of the problem by mounting a tachometer on the motor shaft from which he derived a voltage proportional to the speed of rotation and this voltage was applied to the feedback network together with the error signal. A development of this idea by Williams eliminated the addition of the tachometer to the motor shaft and enabled the necessary feedback voltages to be derived from the armature of the driving motor. Williams named this versatile low-power variable speed drive the *Velodyne*.[4]

The second problem, peculiar to our lock–follow system, concerned the accelerations of the aeroplane into which the device was to be fitted. The fighter aircraft of those days were subject to considerable pitch and yaw motions in level flight and it was essential to discover the exact characteristics of these motions so that the drive system of the lock–follow paraboloid and the control circuits could be designed to take account of them. Our equipment was to be used in Beaufighters and no one at the Bristol Aircraft Company or at the Royal Aircraft Establishment appeared to have this essential data about the pitch and yaw of the plane.

Eventually I was advised that the London Midland and Scottish Railway (LMS) used a device known as a Hallade recorder, for measuring motions of a railway carriage caused by track faults. It seemed possible that this recorder might give the appropriate information for the pitch and yaw motions of a Beaufighter. So, early in October I spent '11½ hours in trains for 20 minutes with Mr Bond in Watford'. W M Bond explained to me the details of the recorder which consisted essentially of pendulums constrained to move in the direction of the appropriate axis. The pendulums were separately connected to needles pressing on recording paper on a rotating drum. With minor adjustments, this equipment seemed suitable for obtaining the measurements I needed and Mr Bond agreed to bring the equipment to Christchurch for flight tests provided I obtained the permission of Sir Harold Hartley, who was then Vice President and Director of Research of the LMS railway. Sir Harold readily agreed and on 7 October we secured the apparatus to the back seat of a Beaufighter which was about twice the distance behind the centre of gravity of the aircraft as the AIF apparatus would be in front. In three flights on that day satisfactory recordings were obtained and were analysed in my report on 29 October.[5] In level flight (appropriate to blind firing) the pitch of the Beaufighter was periodic with average accelerations of $\pm\frac{1}{4}g$, occasionally reaching $\frac{1}{2}g$. In yaw the accelerations were much smaller.

A Blenheim T1939 had been assigned for our use in order to make the first flight trials of this AIF system. This was fitted with the necessary racks and cables at Christchurch while we assembled the complete airborne system in a trailer at Leeson. Metropolitan Vickers in Trafford Park had been allocated the task of making an engineered version of our lock–follow paraboloid. In late August and early September I drove to Manchester on several occasions with the expectation that I would return with this paraboloid, but always there was some minor problem and it was 19 September before this prototype lock–follow scanner arrived at Leeson. By mid-October we had made our final ground tests at Leeson using this MV scanner and 'on Tuesday (14 October 1941) came the great day and we took it [the trailer] to

Christchurch. Blenheim T1939 was well prepared in advance and when we left on Friday afternoon, all was wired and connected and the wheels were turning round'. The AIF system in the Blenheim used an 8λ aperture paraboloid (λ = 9.1 centimetres) which transmitted a 'split-beam' produced by an offset dipole, rotating at about 800 RPM. The receiver output was switched synchronously with the dipole position by means of a commutator and the signal content from the four quadrants was used to drive the mirror system in a direction to equalise the four sets of signals. The signal to be followed was selected by a manually-operated range strobe. The four sets of voltages, representing the signals from the four quadrants, were applied to the grids of the pentodes, phase-advanced in their cathodes. The split field driving motors were in the anode circuits of these valves.

My work which hitherto had been centred on Leeson now moved to the aerodrome at Christchurch and for some time I began to live there instead of in Swanage.

> The plane was ready and all set by Thursday afternoon [21 October] but an engine was u/s. The apparatus was going marvellously on the ground. Wednesday afternoon we took off. Ritson and myself in the back [with the equipment]. After 10 minutes we came back as there was no ASI [Air Speed Indicator] on the plane. Thursday it was u/s and the nose had to be removed for inspecting the scanner and it was Friday before we flew. It worked OK but I felt somehow depressed and disappointed. The cm gear was great on the ground and yet very poor in the air—only 3–3½ miles. Also I didn't think finding [the target] was going to be too easy. Williams and Ritson took it up again in the afternoon with similar results.

That was typical of the many flight tests in the Blenheim during that autumn. In the preliminary search for the target the mirror system, with dipole rotating, automatically swept 120 degrees in 10 seconds and then reversed. The elevation was under the control of the observer, and the sweep could be carried out at any elevation between + 60 degrees and − 20 degrees. The observer had two cathode-ray tubes side by side: on the left a spot tube which indicated the azimuth and elevation position and movements of the mirror, a slight afterglow on this tube conveniently indicated the path just traced by the mirror; on the right a range tube consisted of an ordinary linear time base. When the beam swept over the target an echo appeared on this tube and the observer then switched from 'scan' to 'auto'. In general this operation was too slow and it was necessary for the observer to use an overriding switch in order to make the mirror retrace a few degrees until the echo was visible again. The range strobe was then placed over the echo and the mirror followed automatically. The position of the mirror was taken off by potentiometers and presented both to the pilot, with range as wings, and also to the observer's spot tube. The pilot then flew to centre the spot, the only remaining duty of the observer being to keep the strobe on the echo, and in the final stages to switch from 'coarse' to 'fine' indication so that the pilot's spot tube represented ± 5 degrees total instead of ± 60 degrees.

Sometimes the centimetre equipment would fail and sometimes when it worked we failed to find the target aircraft. It had to be demonstrated to the Duke of Kent and nearly every other visiting VIP and I have few happy diary entries or memories of that autumn of 1941. The constant flying blind in the cramped fuselage of the Blenheim was unpleasant, uncomfortable and sometimes hazardous. Our aerodrome was on the coast—and as Hodgkin had discovered in his early flight tests of AIS that spring, a Blenheim with an odd-looking perspex nose could easily be given uncomfortably close attention as a possible intruder by our Spitfire and Hurricane pilots. The ultimate tragedy was soon to happen.

However, in those autumn days of 1941 we slowly improved the performance of the lock–follow system in Blenheim T1939. By early November about 10 hours of airborne tests had been completed and my report to Dee (dated 4 November 1941) contained the following:

> Although the driving torque available in this first model is insufficient and the mechanism is poor, good following has been obtained. For the low angular velocities encountered in a tail-on approach, the system has held to $\pm \frac{1}{4}^{\circ}$. For angular velocities of about 8° p/sec the lag is approximately $1\frac{1}{2}-2^{\circ}$.

and in respect of the future programme:

> Analysis of the pulse to pulse fading shows that it is necessary to use high speed dipole rotation in order to be able to smooth sufficiently to give accurate following. Similar results can be obtained by using slow speed rotation with many pulses per quadrant, but the time constants involved becomes too great and the recurrence frequency too high. Plans therefore exist for a 6000 r.p.m. model using electronic switching and one pulse per quadrant. The success of the first low speed model however will probably result in a freezing of this simpler design for the first Beaufighter installations. Improved mechanical design and greater driving torque should certainly provide sufficient accuracy for blind firing with fixed forward guns.
>
> For these future models other control circuits and driving mechanisms are under development. The chief competitive scheme uses a magnetic clutch, which gives full driving torque for a few minutes misalignment and is extremely quick acting. [This referred to the Ferranti development.]

Indeed, a month before the first flight of AIF, a decision was made to produce a version of the lock–follow for Beaufighters and in mid-December I was again in Manchester at Metropolitan Vickers 'approving the Type 2 mirror [for the Beaufighter]' and at Ferrantis who were making an engineered version of the control and display units. By Christmas our experimental AIF in Blenheim T1939 was performing well. The lock-on to the target aircraft was extremely good and held securely in rate 2 turns at a range of one to two miles from the target. Acquisition of the target echo became a problem only during continuous banking of the aircraft and then ground returns confused the presentation. However, to overcome this and also when dealing with low-flying targets we developed a technique of flying at low altitude and searching with the paraboloid elevated. By mid-November we were carrying out blind firing trials using a camera gun. The first dozen flights to test the blind firing possibility were made in Blenheim T1939 using a G22 camera gun. There were problems with the camera release mechanism but 81 successful blind fire exposures were made against Blenheim, Gladiator, Battle and Wellington aircraft as targets. The observer made a normal lock–follow interception and then the pilot would close until the wings on his spot indicator showed that he was about 1000 feet from the target. He would then operate the camera gun. Analysis of the records revealed that the statistical probable errors were 0.41 degree in azimuth and 0.32 degree in elevation.[6] Because of the slipstream a dead-astern approach was found to be impracticable and the method adopted was to fly below the target level and bring up the nose of the Blenheim to fire at about five degrees elevation.

The G22 camera gun used in these tests was a single exposure camera and represented instantaneous shots and was not, therefore, simulating combat conditions. Further tests were made with Blenheim T1939 equipped with a G45 cine gun so that the effect of bursts of fire could be assessed. These revealed that with the pilot flying hooded the first few bursts on a given run were more accurate than the later ones. This was attributed to apparent 'tiredness' of the pilot when flying under blind conditions. The equipment was then modified so that the pilot could pre-set his firing button and the camera gun then operated automatically when the target was in line and inside a pre-determined maximum range. An analysis was made of burst durations from less than $\frac{1}{2}$ second to greater than 4 seconds. For 42 per cent of all bursts the errors in both azimuth and elevation were less than ± 0.5 degree. The conclusion of this report[7] was that, with the improvements to be expected in the Beaufighter lock–follow installation, the cannon group of the Beaufighter could be expected to hit the target aircraft for at least 50 per cent of the time using the completely automatic blind firing arrangement.

As the end of 1941 approached I was immersed in the tests of this experimental lock–follow AI in the Blenheim and in the plans to equip a Beaufighter with a prototype AIF in anticipation that a squadron could be operational in 1942. However, in the last days of 1941 my association with the AIF and with these future plans was abruptly terminated. The last references in the whole of my wartime diary to this lock–follow AIF are on 1 January (1942): 'Absolutely heartbreaking to leave AIF and Williams and Ritson are terribly fed up' and, on 10 January: 'Got quickly conditioned to the new job when I knew it was inevitable, but worked overtime with F C W [Williams] to draw out the circuits'.[8] The 'new job' is the subject of the remainder of this book. My absorption with this new task became absolute and apart from completing the reports on the blind firing trials I lost touch with the device that had occupied me throughout 1941. However when writing this account nearly a half century later, I have been able from diverse sources to piece together a summary of the subsequent fate of the lock–follow centimetre AI system.

The prototype installation of the lock–follow AI in a Beaufighter was accomplished and in the autumn of 1942 was flight-tested at Christchurch as AI Mark IX. These airborne trials were satisfactory. With a magnetron producing 50 kW in the pulse a firm lock-on to a target aircraft at a range of five miles was achieved and the lock held securely. Unfortunately the successful tests of this prototype soon led to a tragic accident under the following circumstances. Earlier in 1942 a controversy had developed over the use of 'Window' by Bomber Command. 'Window' was the code-name given to the dropping of a large quantity of metal strips which, if cut to the appropriate size, seriously confused the defensive radars. In essence the controversy was between Bomber Command who wished to deploy the 'window' over Germany in order to confuse the German radars and thus protect the bombers, and Fighter Command who feared that when the Germans discovered the stratagem used by Bomber Command they would use the same technique over England and thereby seriously impede the efficiency of the CH radar chain and the AI in our night fighters.

On 5 May 1942 the Air Ministry decided that Bomber Command would not be allowed to use Window over enemy territory until the effects on our own defensive radars had been thoroughly investigated. Wing Commander Derek Jackson, a distinguished physicist in Oxford who had just completed a successful tour of night operations as a navigator in night fighters equipped with the metre wave AI, was placed in charge of these tests. They were made in collaboration with TRE from the aerodrome at Coltishall in Norfolk. By September 1942 Jackson had confirmed in two reports that our existing night defences were vulnerable to Window and he indicated possible means by which the effects on the AI radars could be reduced, particularly by improving the method of presentation of the radar contacts to the navigator. In this respect the lock–follow AI Mark IX seemed particularly relevant and in December 1942 the prototype AI Mark IX Beaufighter was transferred to Coltishall so that Jackson could assess the performance of the lock–follow radar in the presence of Window.

In the re-organisation at TRE following my removal from the lock–follow AI, A C Downing who had worked with Hodgkin and Edwards on the AI Mark VII was transferred to the lock–follow AI Mark IX programme and he accompanied the prototype to Coltishall. Jackson proceeded to make 13 flights with this AI Mark IX in order to establish whether the lock would hold when the target aircraft was throwing out Window. There were certain problems which Downing proposed to overcome by modifying the radar lock circuits. The modifications were ready for trial on 23 December 1942 with the intention that Jackson and Downing would operate the AI equipment. Shortly before take-off it was discovered that no one was available to launch the Window from the target aircraft so Jackson agreed to do this while Downing tested his modifications to the AI Mark IX in the Beaufighter. The appropriate warning to the Control Station failed to avert the tragedy that ensued. The events have been described by Jackson:[9]

> I heard the control order two Spitfires to scramble: they were to be ready to intercept an unidentified plane. I was somewhat dismayed at hearing this, particularly because the Controller gave no indication of how far away the unidentified aircraft was, and he appeared to be about to ignore my request that the fighters stationed at Coltishall should be kept well away from the two Beaufighters making the Window trial. When we were over the sea I saw one of the Spitfires coming towards us in the most sinister manner, and at once said to Squadron Leader Winward on the intercom that a Spitfire was approaching us in a manner which I did not like. Winward immediately started a very sharp diving turn; almost simultaneously we were hit by cannon fire from the Spitfire. Although neither the pilot nor I were aware of it, this had severed the intercom leads, so each thought the other had been killed, as there was no reply on the intercom. It seemed to me that the only thing for me was to bale out: but the turn was so tight that the g was so strong that I could not reach my parachute; and I had a decidedly unpleasant wait of about ten seconds before what I knew would be the end; but I was in error; a few hundred feet above the water, the Beaufighter levelled out: I saw the fault in the intercom and remedied it just in time to hear Bill Winward say on the R/T '... and he has killed my observer!' Our aeroplane was damaged so we had to return to Coltishall immediately; but as we turned, Bill Winward saw an aeroplane burning, on the surface of the sea; our fears that this was the other Beaufighter and that both the pilot and Dr. Downing had been killed were well founded.'

The death of Downing and the loss of the only prototype AI Mark IX was a severe setback to the lock–follow AI project. It could also have been a severe blow to the use of Window by Bomber Command but at that moment the prototype of the first American centimetre AI, SCR720, arrived and fitted in a Wellington aircraft, was tested against Window by Jackson early in 1943. This American AI, classified in the RAF as AI Mark X used a helical scanner rotating at 360 RPM. The presentation of the echo from the target was on two tubes, one displaying range–azimuth and the other azimuth–elevation. In his tests of the prototype Jackson decided that with this presentation the operator could still make an interception of the target in the presence of Window.† After trials at FIU, AI Mark X fitted in a Mosquito

---

† As a result of Jackson's conclusions Bomber Command was authorised to use Window by the Air Ministry on 16 July 1943. It was first deployed in a raid on Hamburg on 24 July 1943.

made its first operational patrol in May 1943 with Jackson as navigator. Soon, this American AI equipment became available in large quantities and towards the end of 1943 began to supersede AI Mark VIII in the RAF night fighters (by this time largely Mosquitoes). This version remained the standard AI equipment in the RAF until 1957.

These events involving the loss of the prototype lock–follow AI Mark IX on 23 December 1942, the success of the American SCR720 (AI Mark X) against Window and the availability of this equipment from American sources removed the pressure on the lock–follow AI Mark IX. However, work continued on the lock–follow systems in 1943. Burcham was now responsible for the centimetre AI systems and this work was carried out in his group. Another prototype equipment was fitted in a Mosquito aircraft and the FIU tests were satisfactory—although the search facilities were not good enough and the development of a more powerful (200 kilowatts) transmitter was accelerated. Late in 1944 the eventual manufactured form of the lock–follow AI Mark IX was a satisfactory and highly-advanced system capable of operating in the presence of Window and of blind firing. However, I am informed by Burcham[10] that at a meeting on 11 October 1944, which he attended with Dee and Rowe, Derek Jackson criticised AI Mark IX because of bad ground returns. 500 sets of the AI Mark IX equipment had been ordered but the tragic circumstances involving the death of Downing had introduced too much delay for the AI Mark IX to be of significant operational use during the war.

Although my personal records about the lock–follow system terminated at the end of 1941 I discovered, when writing this chapter in 1990, a report of sea trials of the equipment in 1942. According to this report[11] these trials were carried out in a motor gunboat MGB 614 with the scanner mounted on a wooden structure, in the position of the forward gun, so that the height of the scanner above the deck was about 5 feet and above the water line about 15 feet. Tests were made with a target boat crossing the bows of MGB 614 at speeds between 8 and 40 knots and at ranges between 3000 and 9000 feet. The speed of the MGB was varied between 8 and 27 knots. The average lock errors (including those in the rough weather trials) were 0.48 degrees in azimuth and 0.13 degrees in elevation.

Although this report contains detailed tabular matter of each trial and graphs of the following errors there are no indications of the date, place or personnel involved. The date of 6 October 1942 on the drawing office charts implies that the trials must have been carried out sometime in the summer or early autumn of 1942. I am indebted to Dr F J U Ritson (who remained working on AISF after I left the project) for the information that he was the author of this report[12] and that in the spring of 1942, Commander R T Young, in the Experimental Department of *HMS Excellent*, the Gunnery School, proposed that the AISF system could be used to control the 2 pounder (pdr) pom-pom guns in the motor gun boats. A number of trials were carried out by Ritson and F C Williams: (i) with the AISF trailer located on a seaplane slipway at Lee-on-Solent successful trials were made on the tracking of small surface vessels and low flying aircraft; (ii) with the trailer on a rolling platform at Eastney to simulate the motion of a gunboat; (iii) with the AISF equipment in the gunboat MGB 614 as described above in the summer of 1942 at Weymouth.†

For the following information about the sequence of those trials I am indebted to Commander Derek Howse.[13]

> The trials you mention were an attempt to provide MGBs, the equivalent afloat of the night fighter in the air, with radar to permit blind fire by their 2 pdr guns against enemy coastal craft at night. It was designated type 269U... Your TRE drawings show that there must have been trials in MGB 614 before 6 October 1942. The first mention in the ASE monthly report is in January 1943. In March, MGB 680 was fitted with 269U and with type 268, a Canadian-designed X-band set, both of which were demonstrated in July.‡

References

1 Williams F C and Ritson F J U 1946 *Automatic strobes and recurrence frequency selectors* in *J. IEE* **93** pt IIIA 318

2 The Ferranti reports, including the results of the bench tests of their system are included in the papers deposited in the archives at RSRE, Great Malvern, see reference 3

3 The reports and analysis of the high speed fading data on various target aircraft are lodged in the archives of the RSRE library, Great Malvern. *The Fading of 9 cm aircraft echoes Part I*. File D1396 19 August 1941 ACBL/MT *The Fading of 9 cm aircraft echoes Part II Analysis for automatic following*. File D1396 26 September 1941 ACBL/MT

4 Williams F C and Uttley A M 1946 *The Velodyne* in *J. IEE* **93** pt IIIA 317

5 *The random motions of a Beaufighter in normal flight* TRE report 12/85 (File D1387) dated 29 October 1941. The original records of two of the flights and the calibrations are in this file.

6 *Blind Firing on AIS(F)* TRE report 12/94 File D1387 ACBL/EMH 13 January 1942

7 *Blind Firing on AIS(F)—2nd report* TRE report 12/112 File D1759 ACBL/EMH/MB 30 May 1942

† I am indebted to Commander J D Brown, RN, Head of the Naval Historical Branch, for the information that MGB 614 was completed and commissioned on 6 August 1942 and on or about 21 September joined the 18th MGB Flotilla at Weymouth to work up.

‡ I am informed by Commander Howse that the decision was in favour of the Canadian designed X-band Type 268 but that the system did not get to sea until 1945.

8   When writing this chapter in 1990 I discovered amongst my papers several per-
    sonal wartime files containing many details, drawings, circuit diagrams and cor-
    respondence about AIF. These have been deposited in the archive library of
    RSRE Great Malvern.

9   Quoted by Sir Christopher Hartley in the biographical memoir by himself and
    H G Kuhn 1983 *Derek Ainslie Jackson* in *Biog. Mem. R. Soc.* **29** 269–296

10 Burcham/Lovell correspondence January 1991 deposited with the Lovell papers
    in the John Rylands University Library of Manchester.

11 *Report on sea trials with AISF apparatus installed in MGB 614*—undated and
    unsigned TRE report apart from date of 6 October 1942 on TRE drawing
    D100/10698/69. This report has been deposited in the archive library of RSRE,
    Great Malvern.

12 Ritson/Lovell correspondence December 1990 deposited with the Lovell papers
    in the John Rylands University Library of Manchester.

13 Personal communications Lovell/Howse 26 March–20 April 1990, deposited
    with the Lovell papers in the John Rylands University Library of Manchester.

Chapter 8

H_2S—the Background

In his book *One Story of Radar*, A P Rowe wrote that 'On 1 January 1942 Lovell was given charge of the H_2S project and of the sister project, centimetre ASV'.[1] Rowe's statement is only partly correct. At that time centimetre ASV was under development in a separate group in TRE. As will be related it was nine months later when, under most urgent operational circumstances, that development was abandoned and centimetre ASV became the responsibility of my group.

On 29 December 1941 Rowe had summoned me to his office and told me I was to cease working on the lock–follow AI and take charge of the development of a new device, then known as BN (blind navigation) to help Bomber Command. I responded that I did not want to do this because I was anxious to get the AIF system into a Beaufighter. Dee was absent but when he returned on 31 December he said I had to do this and insisted on driving me to the new aerodrome at Hurn, hoping thereby to lessen my resistance. Early the next morning I was again taken to Rowe's office whose patience was rapidly evaporating as he abruptly terminated my further objections by the terse statement that 'there was no alternative'. In that manner I was ordered rather than 'given charge of' a task that I did not want to do and knew little about.

I was soon to be made aware of the circumstances which were as follows. After the evacuation of the BEF from France there was little scope for offensive action against the enemy apart from bombing. Indeed on 15 January 1941 the Air Staff issued a directive† to the C-in-C Bomber Command that the 'sole primary aim of your bomber offensive should be the destruction of Germany's synthetic oil plants...'. However, the U-boat menace to essential shipping was so serious that in March 1941 Churchill issued a directive giving absolute priority to the Battle of the Atlantic, and on 9 March the C-in-C Bomber Command received another directive[3] from the Vice Chief of Air Staff (VCAS): 'I am directed to inform you that the Prime Minister has ruled that for the next four months we should devote our energies to

† Vice Chief of Air Staff to Commander-in-Chief Bomber Command, 15 January 1941. The full directive is in C Webster and N Frankland.[2]

defeating the attempt of the enemy to strangle our food supplies and our connection with the United States...'. A considerable proportion of the available Bomber Command effort then became involved in mine-laying, and in the bombing of the German battleships in dock or under construction. Although a number of German ports suffered some damage there was little effect on industrial and other targets inland.

Four months later, on 9 July 1941 VCAS again gave instructions for a change in the operations of Bomber Command:[4]

> (9 July 1941—VCAS to C-in-C) I am directed to inform you that a comprehensive review of the enemy's present political and economic and military situation discloses that the weakest points in his armour lie in the morale of the civilian population and in his inland transportation system... I am to request that you will direct the main effort of the bomber force, until further instructions, towards dislocating the German transportation system and to destroying the morale of the civil population as a whole and of the industrial workers in particular ....

Churchill's scientific adviser, F A Lindemann†, was a strong supporter of the view that the bomber offensive against the German cities and industrial targets would be an essential prelude to an Allied victory, but according to Churchill[5], he had:

> begun to raise doubts in my mind about the accuracy of our bombing, and in 1941 I authorised his Statistical Department to make an investigation at Bomber Headquarters. The results confirmed our fears. We learnt that although Bomber Command believed they had found the target two-thirds of the crews actually failed to strike within five miles of it. The air photographs showed how little damage was being done. It also appeared that the crews knew this, and were discouraged by the poor results of so much hazard. Unless we could improve on this there did not seem much use in continuing night bombing. On September 3, 1941, I had minuted:
> *Prime Minister to Chief of Air Staff*
> This is a very serious paper [by Lord Cherwell, on the results of our bombing raids on Germany in June and July], and seems to require your most urgent attention. I await your proposals for action.

The investigation referred to by Churchill, on which Cherwell based his paper, was made by D M Butt, a member of the War Cabinet Secretariat. Butt had studied the night photographs taken by Bomber Command crews during raids in June and July 1941. As Churchill stated, this analysis revealed that only one in three of the aircraft bombed within five miles of their target. The failure of Bomber Command was even more serious than this statistic implied, since as Butt pointed out in his report:[6]

† Professor F A Lindemann was raised to the peerage as Lord Cherwell in 1942.

these figures relate only to the aircraft recorded as *attacking* the target. The proportion of the *total sorties* which reached within five miles is less by one-third. Thus, for example, of the total sorties one in five got within 5 miles of the target, i.e. with [in] the 75 square miles surrounding the target.

That was the background to the Prime Minister's minute to the Chief of Air Staff on 3 September 1941, requesting him to give the matter his most urgent attention.

The Sunday Soviet of 26 October 1941

The Sunday meetings in Rowe's office frequently attended by a curious mixture of politicians, high-ranking service officers and scientists were often occasions on which important decisions were taken. The origin of these meetings and of the name 'Sunday Soviets' has been described by Rowe.[7] Few of those Soviets could have had the significance of the one that convened on 26 October 1941. Although the whole of my work for the remainder of the war was to be determined by that meeting I was not a participant. In fact it was customary for Rowe to summon only the senior scientists who had interests related to the subject under discussion. In any event the most significant TRE person present on that day was Dee whose diary entry reads only: 'Big VCCE meeting/Soviet at TRE on how to locate targets'.

This meeting was, perhaps, the most significant outcome of Churchill's 3 September minute to the Chief of Air Staff. No minutes of the Sunday Soviets were kept but it seems probable that it was stimulated by Sir Robert Renwick whom Cherwell had recently arranged to be responsible in the Ministry of Aircraft Production for the coordination of all research, development and production of radar and radio aids in aircraft.† Rowe's reference to the meeting states that: 'Late in October 1941 I held a Sunday Soviet on how to help Bomber Command to bomb unseen targets'.

There were, in fact, two devices already under development in TRE designed to help Bomber Command in this task but they depended on ground transmissions from England and hence were of limited range. The first of these was Gee, the idea of R J Dippy, which at the time of the meeting was being installed in the bomber aircraft. Gee was essentially a navigational system. Three widely-separated transmitters in England transmitted pulsed radar signals simultaneously. The centre station acted as

---

† See D Saward.[8] In the autumn of 1940 Renwick, then Chairman of the London County Electric Supply, had been brought into the Ministry of Aircraft Production by Lord Beaverbrook to coordinate the production of the new four-engined bombers.

the 'master' and the other two as 'slaves'. A receiver in the aircraft measured the difference in time of receipt of the pulses from each slave and the master.

The locus of all points at which a constant time difference is observed between the pulse from the master and one of the slaves is a hyperbola. Special charts in the aircraft of these sets of intersecting hyperbolae enabled the navigator to determine the position of the aircraft. The range is limited by earth curvature because the aircraft receives the direct ray from the transmitter. With the transmitters on the east coast of England, the maximum range of about 350 miles encompassed the Ruhr. At this range the position of the aircraft could be determined within an elliptical area of about six miles by one mile. Gee was first used operationally by the RAF over Germany in March 1942.[9]

The second of these systems, Oboe, was a precision bombing aid. Two radar stations on accurately surveyed positions on the east coast of England controlled the aircraft which carried a beacon transmitter so that it could be guided to fly at a constant distance from one of the ground stations (the 'cat' station). Thus the aircraft flew in a circle with the cat station at its centre. The radius of the circle was chosen so that the aircraft flew over the target. The other ground station (the 'mouse') also measured the range from the beacon carried in the aircraft and transmitted the 'release' signal to the aircraft at the calculated time. As in the case of Gee the range was limited because the aircraft had to observe the direct radar transmission from the ground station. Although the accuracy was very high—at a range of 250 miles an aircraft flying at 30 000 feet could release its bombs to within 120 yards of a selected spot, only one aircraft could be controlled every 10 minutes. However, the use of high flying *Mosquito* aircraft fitted with Oboe to mark the target by dropping flares led to great devastation in the Ruhr. Oboe was first used operationally by the RAF in December 1942.[10]

Although both Gee and Oboe were to prove of great importance to Bomber Command operations, they were ruled out of discussion at the meeting on 26 October 1941 because Cherwell insisted that great ranges of operation were needed to enable Bomber Command to penetrate deep into German territory. According to Rowe:[11]

> We therefore discussed the possibility of self-sufficient equipment in a bomber aircraft which might enable electric power lines to be followed or which might detect towns by virtue of the magnetic field associated with electrical installations. The day ended sadly, for I recall that we went to our homes tired and without an idea.

In the historical context this appears as a somewhat odd comment by Rowe. Although the ground return echoes on the metre wave AI systems had been a serious trouble for air interception, their potentiality for navigation had been recognised before the war by Bowen and his group. Thus, of the early flight trials of the 1½ metre AI from Martlesham Heath, Bowen wrote: [12]

> We had several regular radar targets. The first were the wharves and cranes at Harwich, easily identified soon after take-off from Martlesham. Another was the abandoned airship shed at Pulham near Norwich, a huge metal structure which, on the radar, stood out like a beacon in otherwise featureless terrain.

When I first arrived at Scone airport in September 1939 Bowen said he had ten different airborne radar projects in hand and asked me to look at the files. One of these labelled *Terrain Identification* described a flight he had made in July 1939 across England and Wales completely blind in a Blenheim by comparing the information given by a modified 1½ metre AI with a map. Also in the winter of 1939 at St Athan he had continued his experiments on town identification with metre wave equipment in an Anson, but lack of encouragement and the troubles with the 1½ metre AI caused these experiments to be abandoned.

Hanbury Brown who was involved in several hundred hours of experimental flying with the metre wave radar in 1937–39, refers to a memo [13] dated 17 December 1937 by Bowen and comments:

> AI and ASV were not the only applications of airborne radar which we tested. One of the many things which Taffy Bowen had suggested in 1937 was mapping the ground. In his memo he wrote, '... it is certain that towns and villages would show up from surrounding country, that hedges, trees and possibly railway lines and power lines would also be in evidence'.

As a restrospective judgement on the failure to develop this type of navigational-bombing aid Hanbury Brown makes the appropriate comment:

> Although this work showed that towns could be detected by radar it was before its time, politically and technically. Politically it was a time when most of our politicians were interested in defence and would not have been attracted by a scheme which promised to guide bombers over Europe. Technically it would have been difficult, very likely impossible, to develop a system which would have been of real use to a bomber in finding its target. What was really needed was an airborne radar with a narrow beam which could scan the ground and make a map; this was not possible at metre-waves and had to wait for the development of centimetre-wave radar some years later.

At the time of Rowe's Sunday Soviet on 26 October 1941 Bowen was in the USA and the core of the original metre-wave airborne radar group was dispersed. In the absence of any records of that meeting it must be assumed that there was no one present who was aware of this pioneer work on the detection of ground targets by an airborne radar. In the official British history of the strategic bombing campaign the authors state incorrectly that H_2S was a modified form of ASV, that it had been under development before Cherwell's agitation in 1941 and that:

> Air Chief Marshal Sir Hugh Lloyd informed the authors that he had used it in a Whitley [bomber] and could detect coast lines and the Pennines on an experimental H_2S.[14]

These comments refer to the 1½ metre airborne system developed by Bowen for AI and ASV and these systems were not the forerunners of the wartime H_2S.

In the event it was Dee who, within a few days of the October 1941 Sunday Soviet, revived the idea of town detection in the context of the centimetre radar developments.

References

1 Rowe A P 1948 *One Story of Radar* (Cambridge: Cambridge University Press) p 119
2 Webster Sir Charles and Frankland Noble 1961 *The Strategic Air Offensive against Germany 1939–1945* (London: HMSO) vol 4 pp 132–133
3 Webster Sir Charles and Frankland Noble 1961 *The Strategic Air Offensive against Germany 1939–1945* (London: HMSO) vol 4 pp 133–134
4 Webster Sir Charles and Frankland Noble 1961 *The Strategic Air Offensive against Germany 1939–1945* (London: HMSO) vol 4 p 135 *et seq*
5 Churchill Winston S 1951 *The Second World War* vol IV *The Hinge of Fate* (London: Cassell) p 250
6 *Report by Mr Butt to Bomber Command on his examination of night photographs* 18 August 1941 reprinted in Webster Sir Charles and Frankland Noble 1961 *The Strategic Air Offensive against Germany 1939–1945* (London: HMSO) vol IV appendix 13 pp 205–213
7 Rowe A P 1948 *One Story of Radar* (Cambridge: Cambridge University Press) p 33 *et seq*
8 See Saward Dudley 1984 *'Bomber' Harris* (London: Cassell) p 111
9 For a full description of Gee see Dippy R J 1946 *Gee: a radio navigational aid* in *J. IEE* **93** pt IIIA 468
10 For a full description of Oboe see Jones F E 1946 *Oboe: A precision ground controlled blind-bombing system* in *J. IEE* **93** pt IIIA 496
11 Rowe A P 1948 *One Story of Radar* (Cambridge: Cambridge University Press) p 116
12 Bowen E G 1987 *Radar Days* (Bristol: Adam Hilger) p 48
13 Hanbury Brown R 1991 *Boffin* (Bristol: Adam Hilger)
14 Webster Sir Charles and Frankland Noble 1961 *The Strategic Air Offensive against Germany 1939–1945* (London: HMSO) vol 1 footnote p 248

Chapter 9

The Birth of H₂S— 1 November 1941

At the 26 October 1941 meeting, Cherwell insisted that Bomber Command must have a radar bombing aid self-contained in the aircraft. Although on that afternoon, as Rowe recalled, the meeting ended 'without an idea', the discussion had stimulated Dee to reflect on the possibilities inherent in the centimetre system. Those in the field at Leeson House clearly revealed the radar reflections from the small town of Swanage a few miles distant down the hill. The strong echo from the tin sheet standing out against the ground scatter, observed in the first ground tests of the centimetre system at Worth in August 1940 (Chapter 4) seemed significant. Also several of the centimetre AI systems were now airborne on which coastlines were well-defined. Early in the tests of the spiral scanning AIS system (Chapter 6) submarines had been detected with only the conning tower exposed.

These features led Dee, a day or so after Rowe's meeting, to organise a specific test to discover whether a centimetre radar flying several thousand feet above the ground could distinguish the echoes from a town amongst the echoes scattered from the ground. In Chapters 5 and 6 a description has been given of the two experimental AI systems working in trailers at Leeson in January 1941. As described there, priority was given to the spiral scanner for the first airborne tests in March 1941. The other system used the GEC double paraboloid helical scanner (see Chapter 5) and in August 1941 this was also installed, in Blenheim V6000, for flight trials. As an AI equipment the results were inferior to those obtained with the spiral scanner (AIS) system and during the week following the 26 October meeting Dee asked B J O'Kane and G S Hensby, who were carrying out these tests, to modify the scanner so that it rotated at a constant fixed depression to the horizontal.

Although, as stated in Chapter 8, Rowe's account of the 26 October Sunday Soviet records that, 'The day ended sadly, for I recall that we went to our homes tired and without an idea', it seems that Dee had already envisaged the possible use of a centimetre equipment. On the Friday before that meeting (24 October 1941) O'Kane's diary entry (referring to the tests of the helical scanning system in Blenheim V6000 at Christchurch) reads †

> In the afternoon PID [Dee] came over and stopped our flying as the other 10 cm people wanted to fly and he gave them priority. Actually he had a suggestion which he wanted to discuss with us so there were extenuating circumstances. The suggestion involves a break in the AISH [helical scanning AI] programme for a week or so ...

Philip Dee photographed in his office in the Preston Laboratory of Malvern College, Great Malvern by Dudley Saward autumn 1943. (Reproduced courtesy of Group Captain Saward.)

It seems clear from O'Kane's account that Dee had foreknowledge of the topic for discussion at the meeting on 26 October and had already set in train a test of the idea that an airborne centimetre equipment might be a solution. ‡ O'Kane's comment[3] on the lack of any mention of this during the meeting on 26 October is:

† I am indebted to Dr B J O'Kane for supplying me with a copy of this diary entry.
‡ In a letter[1] to Professor W E Burcham on 7 February 1984 the late Dr John Warren describes a flight with Dr Bernard Kinsey in the Wellington aircraft equipped with the experimental 10 centimetre ASV (see Chapter 18). On the return to base he saw, on the cathode-ray tube, 'sea clutter and also some largish bright patches ... from the Southampton area and some smaller ones inland in areas where there were no trees'. Dr Warren puts the time of this flight 'somewhat uncertainly around October 1941'. However, in Dee's diary[2] he expresses despair about the slow progress with the installation in the Wellington which did not become airborne until December 1941 (see Chapter 18). Although Warren states that he described these observations 'a few days later at a meeting in Lewis' office' and expresses the belief that H_2S arose from then, this was not the case since the O'Kane–Hensby flight (1 November) preceded the Wellington flight by several weeks.

Why nothing was said on 26 October is a mystery. Perhaps Dee feared that the test might be forbidden as an unjustifiable delay in the AI programme. Alternatively he might have wished to have some evidence of a possible solution before raising any hopes.

In the event O'Kane and Hensby visited GEC on 27 October and obtained agreement to interrupt the tests of the helical scanner AI system. The scanner was taken to the TRE workshops on 28 October to be modified for rotation at a fixed angle of depression and on the Saturday (1 November) following Rowe's meeting, O'Kane and Hensby flew this system towards Southampton and, from an altitude of about 5000 feet, the radar echoes from the town were clearly distinguishable from the ground scatter.

Immediately after this encouraging result O'Kane and Hensby made several more flights in Blenheim V6000, carrying a camera to photograph the cathode-ray tube. Flying at 7000 feet over Salisbury Plain the photographs (figure 9.1) showed the responses from the towns of Salisbury and

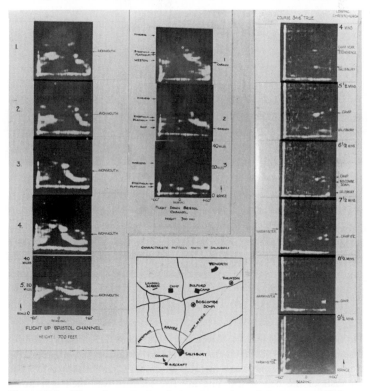

Figure 9.1 The photograph of the CRT when flying over Salisbury Plain and the Bristol Channel, November 1941. The presentation on the CRT is range (vertical) azimuth (horizontal).

Warminster and military encampments. Dee was shown the prints before they were dry and rushed them to Rowe's office. With the wet prints before him on the desk Rowe exclaimed, 'This is the turning point of the war'.[4]

Although these photographs caused considerable excitement in TRE not everyone shared Rowe's immediate opinion. There was no doubt that towns gave clear responses on the cathode-ray tube but so did many other objects —a landing screen near Salisbury gave a particularly strong response, as did the many military camps on Salisbury Plain. The crucial question was whether the radar responses obtained in separate flights were definitely associated with specific ground objects. The news of the initial success quickly reached Cherwell and, impatient with the delay occasioned by the time needed to complete such an assessment and believing that a simple fixed aerial device would give Bomber Command all the essential information, he arranged with the Secretary of State for Air to summon a meeting in order to give a clear directive to TRE. At this meeting on 23 December 1941, after seeing the photographs and hearing the evidence, the Secretary of State

Dr Bernard J O'Kane. In November 1941 when O'Kane and Hensby made the flights in Blenheim V6000 and obtained the photographs shown in figure 9.1 he was a member of the GEC staff attached to TRE for tests of the helical scanning AI. The security was such that Rowe excluded O'Kane from a policy meeting on the implications of figure 9.1 because he was not a member of TRE. (Copyright Marconi Wireless Telegraph Co. Ltd.)

directed that six specific flights should be made immediately to 'determine whether the signals obtained in separate flights could be definitely associated with specific ground objects'. At Cherwell's insistence he also directed that, simultaneously, an assessment should be made of the potential of a simple fixed aerial system.

Scanning versus Split Aerial System Tests

These test flights were made by O'Kane and Hensby early in 1942 and the results were summarised as a Most Secret TRE Report on 23 April.[5] In the modified helical scanning AI system in Blenheim V6000 the 28 inch aperture paraboloid rotated at 300 RPM with the axis inclined about 10 degrees below the horizontal. At the wavelength of 9.1 centimetres the beam was approximately 15 degrees wide and coverage was limited to ±60 degrees in azimuth by the nose structure of the Blenheim. This system used a 'range–azimuth' presentation on the cathode-ray tube with range of the echo as ordinate and azimuth as abscissa. With the Blenheim flying at altitudes of 5000–10000 feet the results were decisive. Southampton, Bournemouth and Wolverhampton were easily detected at ranges up to 35 miles and the angle of approach was immaterial. Aerodromes with hangars gave similar ranges. No response was obtained when flying towards the Black Mountains, but Pontypool was clearly seen. The echoes from Malvern were similarly distinct on the CRT when flying towards the Malvern hills.

During the course of these flights the anxieties about flying the magnetron over enemy territory were already prominent and in view of the ranges on towns given by the equipment in Blenheim V6000 an optimistic assumption was made that the lower-powered klystrons could be used. To test this assumption O'Kane and Hensby made a flight 'with the overall sensitivity reduced to correspond with that obtainable with a klystron. The results showed that range reduction was not sufficient to impair the usefulness of the apparatus'—a conclusion that was soon destined to cause much trouble.

Another critical issue concerned the effect of the altitude of the bomber on the nature of the reflections from ground targets. The new four-engined bombers which were about to come into service operated at altitudes up to 20 000 feet. Blenheim V6000 could not fly above 10 000 feet because the experimental centimetre equipment would flash over. A Beaufighter equipped with AI Mark VII was therefore used to test the nature of the echoes from towns when flying at 20 000 feet. The spiral scanner was depressed so that the fixed beam gave a maximum lobe at a range of 16 miles from an altitude of 20 000 feet. Good signals from towns were seen and it was erroneously concluded that no problems would be encountered with a scanning system operating at that altitude.

In order to test Cherwell's assertion that a simple fixed aerial would be

adequate, two systems were tested. One used two 15 inch diameter paraboloids mounted side by side at a small angle. The other used a single 15 inch
paraboloid with a dipole which could be rotated eccentrically about the axis
of the mirror. Both systems gave a split-beam analogous to that of the metre
wave AI system. For these tests the AI Mark VII electrical units were used
so that the results would be strictly comparable with those observed with the
scanning system in V6000. A high frequency switch was used with a linear
time base in the CRT so that the signals from the split-beam were displayed
to the left and right of the time base. Then, by adjusting the line of flight
so that the signals were equalised, it was anticipated that the pilot could
home on to the ground target. If the aircraft was flying on a course within
± 20 degrees of the target, signals from towns could be seen at a range of
10–15 miles (about a third of the range obtained with the scanning system).
With the single mirror and rotatable offset dipole a crude PPI presentation
was tried as well as the linear time base.

The conclusions were that the fixed split aerial system was 'unlikely
to provide successful bombing of a specified target unless very high
navigational accuracy were available by other methods. Successful selective
bombing within the target area is most improbable. It may however be
definitely asserted that the use of this method would minimise the bombing
of open country'. A further comment was that 'if the use of a klystron were
essential the efficiency of the split aerial system would be severely impaired
as the range obtainable even with the magnetron is already near the
operational minimum'.

Cherwell still had to be convinced that his idea of a quick and simple fixed
aerial system would not be satisfactory. Burcham had this somewhat delicate
assignment and on Saturday 20 April his diary entry reads:[6]

> Took Lindemann [Cherwell] up in N3522 for display of simple H₂S. Not
> much good on RTB but range tube showing quite good DF. Lot of inter-
> ference from blowers. Demonstration probably had the right effect.

The 'right effect' was to persuade Cherwell that there was no quick and
cheap method of helping Bomber Command. It was an idea that we never
contemplated and, as will be seen, when the report was issued, and at the
time of the demonstration to Cherwell, we were heavily committed to the
scanning system.

The Code-Name—H₂S

Various opinions have been published about the origin of the code name
H₂S. R V Jones, at that time Assistant Director of Intelligence and having
close contact with Cherwell, states that when he first heard about the device
the code was TF and that he objected to Cherwell on the grounds that this

would readily be interpreted as 'Town Finding'.[7] According to Jones, Cherwell had said in TRE that it was a stinking matter that the device had not been produced earlier and that the appropriate name H₂S had been suggested in TRE and interpreted as 'Home Sweet Home' to Cherwell. Saward, then Chief Radar Officer to the C-in-C Bomber Command states that the instrument was 'christened "Home Sweet Home", or H₂S for short'.[8] Rowe[9] writes that 'the code name of H₂S, chosen by Lord Cherwell, was associated with his desire that the equipment should be used for homing onto a target: Home Sweet Home'. Flight Lieutenant Dickie (see p 102) who collaborated closely with O'Kane in the early flights of the embryonic H₂S in Blenheim V6000 believes[10] that the correct explanation is that given by Saward in *The Bomber's Eye* where he paraphrases a conversation with Dee when he first came to TRE in December 1941.

> 'Incidentally, why do you call this gadget H₂S?' 'That was Cherwell. He nicknamed it Home Sweet Home, because it enables one to locate and home straight on to the target. Then, when we were trying to find a more brief name, he suggested using the simpler formula of H₂S.'[11]

Of these various opinions my comments are that the device was never known as TF as Jones states. My own files and my diary notes early in 1942 refer to 'BN' (Blind Navigation).† When, in the ordinary course of file circulation, Rowe observed this he insisted that the code H₂S be used as demanded by Cherwell 'because the whole thing was stinking through not having been done years ago', and to remove any suggestion that we were making the equipment for navigation only. Rowe said that Cherwell had demanded that the emphasis be on blind bombing as distinct from navigation. At least this is in accord with Jones' version to the extent that Cherwell said, 'It stinks' but I do not believe that 'Home Sweet Home' originated in TRE as Jones states. I did not hear the 'Home Sweet Home' interpretation of the code H₂S until much later and believe it originated amongst the Bomber Command crews when the device became operational. However, by the time the code H₂S takes the place of BN in my notes there were far more serious issues to worry about than that of the code-name.

References

1 A copy of the Warren/Burcham letter is with the Burcham/Lovell 1990/1991 correspondence deposited with the Lovell papers in the John Rylands University Library of Manchester.

† O'Kane's diary entries of the test flights in the Blenheim early in 1942 refer to BN. The first reference to H₂S in these notes is on 22 April 1942,[12] and in my own diary the first reference to H₂S occurs on 14 April 1942.

2 Dee P I *A personal history of the development of cm. techniques AI, ASV, H₂S and AGLT* (copy is in Lovell TRE record)

3 O'Kane/Lovell correspondence March 1990, deposited with the Lovell papers in the John Rylands University Library of Manchester.

4 Rowe A P 1948 *One Story of Radar* (Cambridge: Cambridge University Press) p 117

5 A copy of this TRE report No 12/106 (File D1738) is inserted in Lovell TRE record at p 4

6 Burcham/Lovell correspondence October 1989, deposited with the Lovell papers in the John Rylands University Library of Manchester.

7 Jones R V 1978 *Most Secret War* (London: Hamish Hamilton) pp 318–319

8 Saward Dudley 1984 *'Bomber' Harris* (London: Cassell) p 136

9 Rowe A P 1948 *One Story of Radar* (Cambridge: Cambridge University Press) pp 116–117

10 Dickie/Lovell correspondence January/February/March 1991, deposited with the Lovell papers in the John Rylands University Library of Manchester.

11 Saward, Dudley 1959 *The Bomber's Eye* (London: Cassell) p 82

12 O'Kane/Lovell, see reference 3, 17 March 1990

Chapter 10

Halifax V9977

At the meeting on 23 December 1941 the Secretary of State not only directed TRE to make the test flights described in Chapter 9, but he also gave instructions that the development of the scanning system should proceed on high priority for use in the heavy four-engined bombers. These heavy bombers, the Stirling, Halifax and Lancaster, were at that time in the process of replacing the two-engined Wellingtons, Whitleys and Hampdens so far used by Bomber Command in operations over enemy territory.†

Immediately following the Secretary of State's meeting, Dee, O'Kane and Hensby visited Boscombe Down (the Aircraft and Armament Experimental Establishment) to inspect these three types of four-engined bombers. They decided that the Halifax appeared to offer more scope and alternative positions for the scanner than the Stirling or Lancaster. In that way the Halifax was chosen for the first trial installation of H_2S. That was the problem placed in my hands on 1 January 1942 when Rowe, although unknown to me, in receipt of the directive from the Secretary of State, ordered me to take charge of the programme. Three days later I made my first visit to the Handley Page works.

In the Blenheim V6000, although the scanner rotated through 360 degrees, the azimuth coverage was limited to ±60 degrees by the nose structure and the presentation on the CRT was range–azimuth. It seemed obvious to us that, for operational purposes, all-round-looking was essential and that the presentation should be on a plan position indicator (PPI) with the time base rotating in synchronism with the scanner so that the navigator was effectively presented with a 'map' of the area over which the aircraft was flying. Another criterion concerned the shape of the beam from the scanner. In the Blenheim there was a 28 inch aperture paraboloid with a 'pencil' beam-width of about 15 degrees. O'Kane and Hensby had obtained ranges of 30 miles or more on towns and other targets. For an all-round-looking scanner giving a radar picture as uniform as possible of the region underneath the bomber we needed a different beam shape. For the best definition we

† When Air Marshal A T Harris became C-in-C Bomber Command towards the end of February 1942 he had only 378 serviceable aircraft with crews, only 69 of which were the new four-engined heavy bombers (see for example Saward).[1]

required a beam as narrow as possible in azimuth but in the 'range' direction we wanted broad coverage. Furthermore, ideally this 'vertical' polar diagram should follow a special law so that the strength of the signal depended only on the target and the aircraft height and not on the range of the target.† However, as will be seen, the 'vertical polar diagram' was to be the cause of endless trouble and it was to be many a long day before we ceased to worry about this refinement.

In our first effort to achieve an appropriate beam shape with a scanner size which we felt could reasonably be fixed underneath the fuselage of the bomber, we decided to try a section of a paraboloid 36 inches in aperture and 18 inches deep. The overall aperture would be close to that of the 28 inch paraboloid carried in Blenheim V6000, and hence the sensitivity should be the same. The greater diameter would give a somewhat improved resolution and we hoped the 18 inch slice of the paraboloid would give an approximation to the beam shape required in the vertical direction. Our idea was to mount this scanner underneath the fuselage in the underturret position of the Halifax protected by a perspex cover (a cupola) 8 feet long, 18 inches deep and 4 feet wide.

On 4 January I arrived at the Handley Page works accompanied by Bob King from TRE who was in charge of fitting equipment in aircraft, and Whitaker from Nash & Thompson who were producing the spiral scanners for AI Mark VII and who were to be pressed into manufacturing the scanner for the Halifax installation. Handley Page must have received a firm instruction from the Air Ministry or MAP because it was he and his chief design team who met us on that Monday morning. Naturally he was outraged at our audacious request to remove the underturret and mount this large blister underneath the fuselage. His bomber had been under design and development since 1936 with a view to carrying a greater bomb-load at higher altitude and greater speed than any previous airplane. He said our device would ruin the performance of his bomber and force a reduction of the bomb-load. Our response that it was preferable to place a few bombs in the right place rather than drop the maximum load on open country had little effect because we were under orders not to explain that we were developing a new type of blind bombing equipment.

This was the first of many occasions on which no action would have been taken had not the Prime Minister been urged by Cherwell to give the highest priority to H$_2$S. Cherwell had promised us there would be no delay and there was none. On 27 March the Halifax V9977 with the perspex cupola in the underturret position landed at Hurn airport (figure 10.1) 'Saturday March 28th ... After being expected every day this week, the Halifax came yesterday and we were there to receive it and get things moving'. The 'get

---

† If α is the angle to the vertical then it can be shown that the polar diagram should follow a cosecant α law.

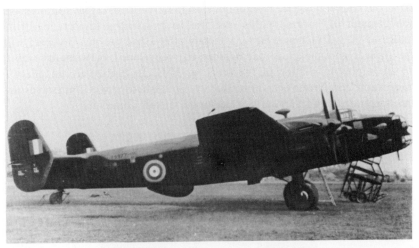

Figure 10.1 The Halifax V9977 at Hurn airport, March 1942. The perspex cupola housing the H_2S scanner can be seen immediately beneath the RAF insignia.

things moving' referred to the installation in the Halifax of the experimental system we had assembled at Leeson since January. The major centimetric units were the magnetron transmit–receive systems then being produced for AI Mark VII (see Chapter 6). The essential new developments needed were the circuits for the synchronisation of the rotating time base of the plan position indicator with the scanner rotation. An important issue that had to be settled at an early stage was that of the rotation rate of the scanner; a consideration that involved several factors, such as pulse width, pulse recurrence frequency, maximum range needed, and maximum available mean power—issues already discussed in Chapter 6 in relation to AI. In the PPI the signals were applied to the grid of the CRT so as to intensify the spot and a long afterglow tube was used with the afterglow slightly longer than the period of rotation of the scanner. Some of the complex issues involved have been described by C J Carter,[2] but in the early months of 1942 we were constrained by the available AI Mark VII units and were concerned essentially with reproducing the Blenheim V6000 results in the Halifax with the all-round-looking scanner and plan position indicator. In the event we decided on a low scanner rotation rate of about 80 RPM.

In October 1941 an RDF Committee was formed in the Air Ministry, initially to deal with the introduction of Gee into Bomber Command. Renwick was appointed Chairman and in *The Bomber's Eye*[3] Dudley Saward has described his interview with Renwick which led to his posting to the Bomber Command Headquarters at High Wycombe as a Wing Commander in charge of a newly created RDF department. Although the need for the rapid introduction of Gee to Bomber Command was the initial stimulus, the appointment of Saward to this post was to be of great importance to the future of H_2S. I first met Saward shortly after I had become involved with the project and the close partnership that developed between us was to be an important matter in the story of H_2S. In particular, an effective integration and liaison was immediately established between ourselves and the operational staff and service personnel of the Command.

Group Captain Dudley Saward (1943). (Reproduced courtesy of Group Captain Saward.)

A Navigation Branch already existed at Bomber Command headquarters under the Chief Navigation Officer, Wing Commander G I L Saye. It was a member of his staff, Flight Lieutenant John Dickie a navigator of considerable operational experience already in close touch with TRE over the development of Gee, who first brought Saward to TRE.† Dickie was Saye's Staff Officer responsible for all operational aspects of radar navigation devices and on 5 January 1942 he tested the experimental system in Blenheim V6000. His report was enthusiastic and his suggestions at that time and in the future were a significant factor that determined the operational usefulness of H_2S. Then, on 20 January, Saward, well aware of our

† Dickie[4] brought Saward to TRE in December 1941 but my initial contact with Saward occurred early in 1942[5].

urgent need for help, posted three RAF flight mechanics to TRE to help with the building and assembly of the units needed for the Halifax. Corporals Dale, Hinckley and Calcutt were the first of many service personnel who were soon to work in my group and exert a critical influence on the progress of the device. It was fortunate that the day before their arrival we were able to transfer from the stables and the trailers to a wooden hut that had been built adjacent to the stable block in the field at Leeson. It was a luxury that we were not to enjoy for long. Saward also sent three RAF officers but, whereas Dale, Hinckley and Calcutt immediately grasped the soldering irons, the employment of the officers was not so simple, 'but we sent them on a course at Forres† to ease the situation'.

The Halifax that had arrived at Hurn on 27 March had to be cabled and fitted with racks to take the units which had been assembled at Leeson. By 16 April the Nash & Thompson hydraulic scanner had been fitted into the cupola and all our units installed. By the evening we were ready for the first test flight—but nothing worked, there was no power from the 80 volt alternator‡ on the outer starboard engine. With this problem rectified we tried again on the morning of 17 April. The equipment worked but very poorly. At 8000 feet altitude the range on towns was only four to five miles and the gaps and fades in the PPI picture implied that there must be something horribly wrong with the polar diagram of the scanner. My brief diary note typifies the scene in which we were to work for many months:

> ... we had DCAS [Deputy Chief of Air Staff], DOR [director of operational research], D of R [director of radar], DD of Ops [deputy director of operations] and various others down on the H_2S. Halifax rushed to be ready for them. Flew OK but too many loose ends to demonstrate properly§.

The Political Pressures

The assembly of such high-ranking officers to witness the first airborne tests of an experimental system may seem remarkable. On that April morning it certainly seemed so to me and my diary entry records my irritation with Dee whom I thought should have had the scientific sense and power to prevent this. However, I did not know the acute political pressures that had recently developed and which were to inhibit the appropriate research investigation of how to obtain a satisfactory H_2S picture in a high-flying bomber aircraft.

† A school in Swanage which had been taken over by TRE and where J A Ratcliffe and L G H Huxley set up the TRE teaching school and the Post-Design Services (PDS) organisation (see Chapter 6 for an explanation of the PDS organisation).
‡ The evolution of the 80 volt alternator as the primary power source for airborne radar equipment has been described by E G Bowen.[6]
§ According to O'Kane's diary in the afternoon he gave another demonstration to DCAS in the Blenheim which created a good impression.

These pressures had their origin in Cherwell. In Chapter 8 I have referred to the intervention of Cherwell with the Prime Minister over the question of our night bombing accuracy and the sequence that led to the exploratory flights in Blenheim V6000 in November 1941. Early in 1942 Cherwell sent a further memorandum to the Prime Minister in which he argued that within the next one and a half years Bomber Command could de-house the majority of people living in Germany's 58 largest cities—and he concluded that this was the best strategy to help the Russians. Cherwell had earlier agreed that an assessment of the damage that Germany might suffer from bombing could be made by a survey of the damage inflicted by the German bombing of Birmingham and Hull. Zuckerman and Bernal were in charge of this survey and Cherwell's memorandum to the Prime Minister had been based on these data. Zuckerman[7] has described how he was asked to provide information for Cherwell's use during the course of the survey. Although the final report of the survey was not complete until 8 April, Cherwell had already submitted his memorandum to the Prime Minister on 30 March. The Zuckerman–Bernal report on their survey of the bombing of Hull and Birmingham stated that:

> ... in neither town was there any evidence of panic resulting either from a series of raids or from a single raid

and in the summary of conclusions they specifically stated:

> There is no evidence of breakdown of morale for the intensities of the raids experienced by Hull or Birmingham (maximum intensity of bombing 40 tons per square mile).

In his account of these events Zuckerman states that this conclusion:

> ... was almost the reverse of what the Prof. [Cherwell] stated. [that is in his memorandum to the Prime Minister]

The Prime Minister circulated Cherwell's memorandum and this quickly led to a violent dispute between Cherwell on the one hand and Blackett, Tizard and others about the wisdom of the bombing policy which Cherwell was advocating. Blackett had recently formed the operational research group in the Admiralty and when asked to comment on Cherwell's paper his response was that:

> ... the *method* of calculation used was correct in principle but that the actual numerical answer, that of the number of houses which could be destroyed within 18 months, was 6 times too high. The main mistake in the calculation was that it was assumed that all bombers which would be delivered from the factories in the next 18 months would in the same period have dropped all their bombs on Germany ...

Tizard had also independently come to the same conclusion that Cherwell's paper exaggerated the probable effect of the bombing offensive between April 1942 and October 1943 by a factor of five to one. To the dismay of Blackett and Tizard, Cherwell's paper was accepted.

> The Air Ministry agreed with the Cabinet Office paper and the policy of making a major contribution to the Allied war effort, until the autumn of 1943, the de-housing by bombing of the German working-class population, with the object of lowering her morale and will to fight, became official British policy.[8]

Over 30 years later Zuckerman[9] made the following comment:

> ... it was not the full report that we [Zuckerman and Bernal] had submitted but the short paper Cherwell based on it which formed the basis of the argument. Cherwell had extrapolated from our facts in far more vigorous fashion than we would have done.

That was the background, of which I was entirely unaware in April 1942, that led to the intense political pressure for the crude form of H₂S then existing, first manifest on the occasion of the initial flight of the Halifax on 17 April.

The Klystron–Magnetron Dispute

During the first flight of the Halifax on 17 April, ranges of only four to five miles were obtained on towns compared with the 30 to 35 miles with the equipment in Blenheim V6000. There was no significant improvement in subsequent test flights and the overall results were very poor with gaps and fades occurring on the PPI screen. This was disturbing and depressing since we were using the AI Mark VII magnetron and had to face the unpleasant fact that authority for flying the magnetron over enemy territory might not be given. In fact, at the Secretary of State's meeting on 23 December 1941, opinion was heavily against the use of the magnetron and the directive was to develop H₂S using a klystron. The ability to generate considerable powers in the centimetre waveband was regarded as a most important Allied secret. If the magnetron was flown in a bomber over Germany the inevitable loss of a bomber containing this equipment would soon reveal the secret to the enemy. On the other hand the principles of the klystron were well-known. Furthermore it was destructable by an explosive charge and hence the official directive was to develop H₂S using the klystron and not the magnetron.

The klystron, particularly in the reflex form, (see Chapter 6) was used in the AI Mark VII as a local oscillator and no forms then existed which were capable of producing pulsed peak powers of more than a few kilowatts.

In the report on the test flights in Blenheim V6000 (see Chapter 9) O'Kane and Hensby had reduced the sensitivity to what they believed might correspond to the use of a klystron and had concluded that 'the range reduction was not sufficient to impair the usefulness of the apparatus'. However, that conclusion was based on a magnetron equipment giving ranges of 30–35 miles on towns and now we were obtaining ranges of five miles and less in the Halifax. Clearly, replacing the magnetron by the lower-powered klystron would be useless.

During the early visits of Saward and Dickie to Leeson they had summarised the operational requirements for H_2S—a range of 30 miles at 18 000 feet altitude for efficient navigation and target approach, and a clearly defined picture at short range for bombing. Even on the basis of the Blenheim V6000 results no one in TRE acquainted with the issues believed a klystron version of H_2S could be developed to meet these requirements. Nevertheless, the official Air Ministry/MAP directive was that the klystron be used as the transmitter. EMI had been chosen to manufacture the equipment and the contracts with that firm had specified the klystron.

In our meetings with Saward we devised a compromise that at least enabled us to proceed with the development of the magnetron version. Early in January 1942 Saward had instigated the formation in Bomber Command of a special Flight (at first Flight No 1418 and later to be known as BDU— Bomber Development Unit)† for the purpose of developing operational techniques for the use of the various radar aids. It was agreed with Saward that we would continue to develop the magnetron version in Halifax V9977 and that a second Halifax which we had requested in our 4 January meetings with Handley Page would be equipped with the EMI klystron version. Both these aircraft would then go to the Flight 1418 (BDU) for comparison in Service trials and experimental assessment. The official request to Air Ministry was made by Bomber Command and approval given on 26 March—

---

† See Saward.[10] Early in December 1941 Saward and Saye had advised Saundby, the Deputy C-in-C Bomber Command, to establish a development unit in connection with the operational use of Gee. On 14 December the TR1335 Development Unit was formed [TR1335 was the Service number for Gee] and on 5 January 1942, it was officially established as Flight 1418 under the command of Squadron Leader O R Donaldson. I am informed by Dickie[4] that previous to 14 December 1941 he had assembled a team of tour-expired colleagues from 115 Squadron at Marham to conduct the Gee operational trials in the late summer of 1941 and that this formed the core of the TR1335 Development Unit established on 14 December 1941 and hence was the forerunner of Flight 1418 (5 January 1942) and of BDU (summer 1942). BDU is variously referred to as Bombing Development Unit and Bomber Development Unit in the published records (including those of Saward). Saward's most recent recollection[11] is that it should be Bombing Development Unit, but Dickie[4] states that it was an abbreviation for Bomber Command Development Unit and should therefore be Bomber Development Unit.

the day before the first Halifax arrived at Hurn. The wisdom of proceeding with the development of the magnetron version was soon to be justified: however, tragic circumstances were to intervene and nullify the plans for the comparative assessments at BDU.

EMI and Halifax R9490

The strength of the political pressures for H_2S and Cherwell's belief in its efficiency is nowhere better illustrated than in the assumption that an immediate transfer could be made of the equipment in Blenheim V6000 to the four-engined bombers. Shortly after the Secretary of State's meeting on 23 December 1941 contracts were placed with EMI for the *manufacture* of 50 complete H_2S units. EMI placed their best engineers on the programme headed by A D Blumlein, a pioneer in the development of television and widely regarded as one of the best electronic engineers in the country. It had been agreed that Curran should be responsible for our main liaison with EMI. O'Kane had by far the most experience of operating airborne centimetre equipment in its embryonic H_2S mode, first in the Blenheim V6000 and next in the Halifax. With O'Kane giving advice from his experience of flying the system and with Curran establishing the close TRE/EMI link we did not, at first, anticipate any serious problems. Indeed, if it had been only a matter of re-designing the AI Mark VII units using the magnetron or developing the additional electronic and control units required for the specific H_2S application the task of manufacturing 50 complete units would not have been a difficult matter for a firm with the great experience of EMI. However, as mentioned in the previous section, this was not the case. The official directive to ourselves and EMI was that the klystron should be used as the transmitter, and—quite apart from the problem of producing a klystron with adequate pulsed power—this naturally reacted throughout the entire electronic system.

At our meeting with Handley Page on 4 January 1942 we had asked for two Halifax bombers to be fitted with the perspex cupola in the underturret position. As already described the first of these arrived at Hurn airport on 27 March and, after being fitted with a Nash & Thompson hydraulic scanner and our experimental magnetron equipment, became airborne in mid-April. The second Halifax R9490 with the perspex cupola arrived at Hurn on 12 April.

During my association with the lock–follow AI in 1941 I had close contacts both with Nash & Thompson and with Metropolitan Vickers on the question of the provision of the lock–follow paraboloid. In view of the wide range of uncertainties about the nature of H_2S in the four-engined bombers I thought it would be expedient to continue these associations and obtain both a hydraulic and electrically-operated scanner. These were produced, the hydraulic scanner was fitted in the Halifax V9977 and the Metropolitan Vickers electrically-driven scanner with the same size and shaped truncated paraboloid was fitted in Halifax R9490. During those early months of 1942 we were therefore in continuous discussion with EMI at Hayes and with these two firms who were making the prototype scanners—one at Surbiton and the other in Manchester. The EMI team wisely decided to build their development model and get it operating in a coach at Hayes. They drove this coach with the klystron version of H_2S to Hurn airport on 14 May. This equipment was soon transferred from the coach to the Halifax R9490. In the meantime we had continued our efforts to improve the magnetron H_2S in V9977 but made little progress. Then on 22 May I 'flew in the Halifax. Very depressing; picture is extremely bad at the moment.' It was therefore scarcely a surprise that when R9490 became airborne with the EMI klystron H_2S on 2 June we had 'the same old succession of aerial troubles'—that is gaps and fades and poor range.

It was hard to see at this time how a klystron version of H_2S with this poor performance could be of much help to Bomber Command and although Saward's faith in the potentialities of the system never wavered, others in Bomber Command were cautious. For example, Group Captain Saye complained bitterly about the 'snowstorm' which is all he claimed to have seen on the PPI and his letter of complaint to the Air Ministry was responsible for a meeting summoned by the Assistant Chief of Air Staff (ACAS) on 19 May. The Air Staff directive emerging from this meeting revealed how urgently Bomber Command needed even this imperfect system:

2 (a)   That the system should be accurate enough to guarantee that bombs would fall within an industrial or other area selected as a target.

   (b)   That the Air Staff would be satisfied in the first instance if the range of the device enabled the aircraft to home on a built-up area from 15 miles at 15,000 ft.

3   Subject to there being no delay or interference with the development of the equipment and its introduction into the Service in a form which will fulfil this aim, it was agreed that details in design to enable it to be used as a navigational aid to determine a specific area or target could be incorporated during the later stages of development and operational trial.

At the time of the ACAS meeting on 19 May which led to this directive we had two Halifax bombers equipped with the experimental H_2S systems at Hurn airport and, with the resources of EMI and their outstanding electronics team and manufacturing experience, there was reasonable hope that this modified target of the Air Staff—15 miles from 15 000 feet—could soon be achieved. This hope was soon quashed by two events, one disturbing and the other tragic.

References

1 Saward Dudley 1984 *'Bomber' Harris* (London: Cassell) pp 114–115
2 Carter C J 1946 *H_2S: an airborne radar navigation and bombing aid* in *J. IEE* **93** pt IIIA 449
3 Saward Dudley 1959 *The Bomber's Eye* (London: Cassell) Chapter 8
4 Dickie/Lovell correspondence January/February 1991, deposited with the Lovell papers in the John Rylands University Library of Manchester.
5 Saward Dudley 1959 *The Bomber's Eye* (London: Cassell) pp 95–96
6 Bowen E G 1987 *Radar Days* (Bristol: Adam Hilger) pp 61–64
7 Zuckerman Solly 1978 *From Apes to Warlords* (London: Hamish Hamilton) Chapter 7
8 See Lovell Bernard 1975 *P M S Blackett: a biographical memoir* in *Biog. Mem. R. Soc.* **21** pp 1–115, for a further description of this dispute and for the original references.
9 Zuckerman Solly 1975 *Scientific advice during and since World War II* in *Proc. R. Soc. A* **342** 471
10 Saward Dudley 1959 *The Bomber's Eye* (London: Cassell) p 73
11 Saward/Lovell correspondence 1990/1991 deposited with the Lovell papers in the John Rylands University Library of Manchester.

Chapter 11

Life in Swanage 1940–42

by Joyce Lovell

Towards the end of May 1940 I was able to join my husband in Swanage—a small seaside town in Dorset. There were plenty of unfurnished houses to let and very quickly we settled in our new home, with our daughter Susan who was one and a half years old. Beautiful countryside surrounded us, and after the bleakness of aerodromes in Scotland and Wales the whole group was thankful to find itself in a sunny, warm place at last. Sadly I did not keep a diary during those years in the early part of the war, but I did write regularly to my parents who lived in Bath and the preservation of these letters remind me of the way life was then. Just for a few halcyon weeks life remained calm and bright. We still had a little petrol and could explore beautiful beaches and parts of the Dorset coastline. So far there was little in the way of sea defences and no barbed wire restricted our access to the beaches. I record in my letters how we enjoyed our lunches of fresh fish. 'The fishmonger seemed quite pained when I failed to order either crab or lobster'. A farmer called at the house twice a week with a large flat basket in which were arranged dressed chicken, brown eggs warm from the nest, dewy-fresh soft fruit and early vegetables. We had a rough patch of ground behind our house. We were the first people to live there and very soon had a vegetable plot, double-dug, ready for planting and we got fine crops.

We lacked nothing in the way of good food and found Swanage people full of goodwill towards us from the very beginning. One could say that they could not do enough and, after those wretched winter months, our immediate future seemed rosy. It seemed too good to be true. It was. Before our first month had come to an end, everything had changed. With little warning we found the war on our doorstep. By mid-May, to our horror and incredulity, Hitler's armies began their advance. They had entered Holland and Belgium capitulated; the British Expeditionary Force was surrounded, and on the beaches of Dunkirk was left to what appeared to be a hopeless fate. Before long some of us were driving to a high point from whence it was possible to watch the assembling of a unique fleet of small ships, pleasure boats, and fishing vessels combining to set off to the rescue of that army left in such dire peril. They crossed the Channel, defenceless, not just once but again in many cases. That story of rescue became a legend. With the boats setting out so close to us, from Poole Harbour, it marked an historic moment in our lives. We had to wake up and realize that now we were looking over the sparkling blue waters towards our enemies all along the coast. Country after country was ranged against us, so that now we must stand alone.

In government offices the unbelievable had been foreseen, and it was only a matter of days before the leaflets were being pushed through the letterbox, and the Prime Minister was telling us exactly what we had to do. The Ministry of Information sent us, not suggestions but commands. I am glad I kept these leaflets carefully with other records in a scrapbook. The words 'your countrymen' appeared frequently and the assumption from the beginning was that we should all want to play our part worthily. I remember thinking, listening to Churchill's commands, 'How strange, that is exactly how I do feel'. The 'us' of a household had widened, and we began to be transformed into countrymen. When France fell it seemed as if nothing more awful could happen. At such moments it is sometimes helpful to settle to some small humdrum task, so I started to clear out unwanted things from the cupboard under the stairs. That was popularly believed to be the safest place during an air raid, in the absence of a proper shelter. I arranged a camp-bed, blankets and gas masks, as well as a few iron rations, toys and story books.

Susan was not yet two years old, but she loved the cupboard, and when the 'tip and run' raids began she quickly settled for a cramped few hours, whilst we listened to the planes flying over. 'That's one of ours' we said. 'No, it isn't.' I wonder how often we were right? After the all-clear sounded we continued peeling the potatoes. We were not really ready for it when we were awoken one morning by the crack of machine-gun fire. Our grocer's shop was close to the small railway station, so we should have been prepared. The raiders were able to retreat very quickly over the coast. Soon afterwards, a notice outside the shop read 'Business as usual', but the shop was shattered and a trembling friend stood there pointing out 'what I *thought* was tomato ketchup, but it wasn't'.

We listened for the wailing sirens from then on, especially when we had children in our care. The increased fortification of the beach began. There were pillboxes at both ends, and long coils of barbed wire covered the whole area. Mines were laid. No longer was it bucket and spade country as in the days of my youth in the twenties. Instead, buckets of sand and long-handled spades stood at our gates, ready to deal with any incendiary bombs which might fall. A notice, STIRRUP PUMP KEPT HERE, was propped up in our windows. I found it in my scrapbook. We learnt how to smother that sort of bomb correctly. We took our turn at duty in the street once a week, and our warden dealt with any problems. I think exemption from duty could be claimed in cases of caring for the very young and very old. I do not remember that many people claimed this exemption. A friend who served in the Royal Navy during World War I divided the night into naval watches, which were scrupulously observed by his wife and older children. One was issued with a steel helmet when on duty. At first things were quiet and we listened to the planes passing overhead bound for Bristol or the Midlands. Southampton was frequently attacked and lit up by blazing fires. Troops guarded Swanage beach in case it offered an attractive landing-site for the enemy— wide flat, shallow and sandy. After a while, rather amazingly, local people only, on showing their identity cards were allowed limited access for one hour every day. We could then enjoy the delights of a paddle or bathe in those sparklingly-clear seas. Through that narrow gap in the barbed wire I not only went on every opportunity, but took a troop of friends' young children with me. The children walked into the sea towards the Isle of Wight as soon as the soldier on duty let us through. The problem arose when the wailing of the air raid siren was heard. I kept a push-chair close to where the soldier guarded the exit, and we all had to run for it. It took a good ten minutes to reach home, and it was all uphill, steep too, I crammed all the little ones in and the older children ran beside and helped push. By the time we reached the house and crowded into the cupboard we were breathless, after all it was my responsibility to keep them all safe. Occasionally the beach was entirely shut off, and those were sad days.

Before long the first leaflet was followed by another, 'Beating the invader'. Years later one of my sons mocked at the idea that a person like me would ever keep a stick or broom or pitchfork behind the door to attack an invader. But the mildest mother in the animal kingdom rivals the tigress if her young are threatened. That is the way I felt then, anyhow, in September 1940. I took good care of the young, but as it never came to an affray how shall I ever know how I would have reacted? Things have not seemed so simple since then.

Rationing began, and although the amounts of sugar, fat, eggs and meat were modest, at least it was fair and our coupons were always honoured, which means that Lord Woolton (Minister of Food) should be congratulated, especially as time went on and shipping losses increased. We could get priority foods for expectant mothers and young children, such as milk and orange juice. The juice came in a bottle which one diluted. There was a certain amount of queueing but I do not remember much in Swanage—that was to come later. If one stayed away from home, even for a few nights, one always gave up one's normal coupons and exchanged them at the Food Office for an emergency card which one presented to one's hostess and so saved the embarrassment of eating someone else's rations.

None of the wives in our group had cars, so we tried to share duties between us in order to give one another a chance to organise ourselves efficiently. One of us did all the shopping, and cooked meals for two or three families together, pooling rations. In this way we could do other jobs needed

Issued by the Ministry of Information in co-operation with the War Office and the Ministry of Home Security

Beating the INVADER

A MESSAGE FROM THE PRIME MINISTER

IF invasion comes, everyone—young or old, men and women—will be eager to play their part worthily. By far the greater part of the country will not be immediately involved. Even along our coasts, the greater part will remain unaffected. But where the enemy lands, or tries to land, there will be most violent fighting. Not only will there be the battles when the enemy tries to come ashore, but afterwards there will fall upon his lodgments very heavy British counter-attacks, and all the time the lodgments will be under the heaviest attack by British bombers. The fewer civilians or non-combatants in these areas, the better—apart from essential workers who must remain. So if you are advised by the authorities to leave the place where you live, it is your duty to go elsewhere when you are told to leave. When the attack begins, it will be too late to go ; and, unless you receive definite instructions to move, your duty then will be to stay where you are. You will have to get into the safest place you can find, and stay there until the battle is over. For all of you then the order and the duty will be : " STAND FIRM ".

This also applies to people inland if any considerable number of parachutists or air-borne troops are landed in their neighbourhood. Above all, they must not cumber the roads. Like their fellow-countrymen on the coasts, they must " STAND FIRM ". The Home Guard, supported by strong mobile columns wherever the enemy's numbers require it, will immediately come to grips with the invaders, and there is little doubt will soon destroy them.

Throughout the rest of the country where there is no fighting going on and no close cannon fire or rifle fire can be heard, everyone will govern his conduct by the second great order and duty, namely, " CARRY ON ". It may easily be some weeks before the invader has been totally destroyed, that is to say, killed or captured to the last man who has landed on our shores. Meanwhile, all work must be continued to the utmost, and no time lost.

The following notes have been prepared to tell everyone in rather more detail what to do, and they should be carefully studied. Each man and woman should think out a clear plan of personal action in accordance with the general scheme.

Winston S. Churchill

STAND FIRM

1. What do I do if fighting breaks out in my neighbourhood?

Keep indoors or in your shelter until the battle is over. If you can have a trench ready in your garden or field, so much the better. You may want to use it for protection if your house is damaged. But if you are at work, or if you have special orders, carry on as long as possible and only take cover when danger approaches. If you are on your way to work, finish your journey if you can.

If you see an enemy tank, or a few enemy soldiers, do not assume that the enemy are in control of the area. What you have seen may be a party sent on in advance, or stragglers from the main body who can easily be rounded up.

Poster 1

at that time. I ran a school of about 15 children, five and six years old, as I was a trained teacher anyhow. There was a flat roof on our garage, so that is where we worked in what seemed in retrospect to be endless sunshine. Writing paper became very scarce but I made quite a lot of material for the learning of the three Rs from our own stores. The children came for two hours each morning and brought a shilling (5p nowadays) for each day they were present. It was not very much, but it was enough for me to contribute each month to the comforts parcels for the Merchant Navy. This was a wartime service I venerated above all others. Picking up the survivors of those ships that had been torpedoed, the idea was to hand them immediately a parcel containing a pair of thick dry socks, a bottle of rum, chocolate and cigarettes. Not much, but something. With each sum of money I forwarded an official label, filled in by the best pupil in the class. What an honour to be chosen to address those labels 'Peter Brown, a member of Mrs Lovell's school'. I sometimes wondered if I should ever meet one of those recipients of a comforts parcel. A friend in the Merchant Navy who was torpedoed later told me how he welcomed the parcel.

The possibility of invasion began to fill our minds. A warning sign would be the ringing of church bells. Our instructions from the Ministry of Information which fell on the doormat are shown in Posters 2 and 3.

Poster 2

done just what the enemy wanted you to do.

How shall I prepare to stay put?

Make ready your air-raid shelter; if you have no shelter
prepare one. Advice can be obtained from your local Air Raid
Warden or in " Your Home as an Air-raid Shelter ", the
Government booklet which tells you how to prepare a shelter
in your house that will be strong enough to protect you against
stray shots and falling metal. If you can have a trench ready in
your garden or field, so much the better, especially if you live
where there is likely to be danger from shell fire.

How can I help?

You can help by setting a good example to others. Civilians
who try to join in the fight are more likely to get in the way
than to help. The defeat of an enemy attack is the task of the
armed forces which include the Home Guard, so if you wish to
fight enrol in the Home Guard. If there is no vacancy for you at
the moment register your name for enrolment and you will be
called upon as soon as the Army is ready to employ you. For
those who cannot join there are many ways in which the Military
and Home Guard may need your help in their preparations.
Find out what you can do to help in any local defence work that
is going on, and be ready to turn your hand to anything if asked
by the Military or Home Guard to do so.

If you are responsible for the safety of a factory or some other
important building, get in touch with the nearest military
authority. You will then be told how your defence should fit
in with the military organisation and plans.

What shall I do if the Invader comes my way?

If fighting by organised forces is going on in your district and
you have no special duties elsewhere, go to your shelter and
stay there till the battle is past. Do not attempt to join in
the fight. Behave as if an air-raid were going on. The enemy
will seldom turn aside to attack separate houses.

But if small parties are going about threatening persons and
property in an area not under enemy control and come your
way, you have the right of every man and woman to do what
you can to protect yourself, your family and your home.

Stay put.

It's easy to say. When the time comes it may be hard to do.
But you have got to do it; and in doing it you will be fighting
Britain's battle as bravely as a soldier.

(Printed in England)

Poster 3

This was not simply advice, this was an order from the Government, and
it was a help to be told what to do when events became threatening. We listened to news of the utmost gravity, but sticking to the principle of sharing
our chores we were able to enjoy some diversions. We played tennis, frequently went to the cinema—I suppose now those films would be called pro-
paganda but they helped like the beat of a drum. We went to dances at the
Grosvenor Hotel which often served as an Officers' Mess, and I remember
one evening being told by a friend that there was a message 'for all the girls'
from the Commanding Officer. A group of young airmen had been shot
down in flames during a recent battle. Their faces were very severely dis-
figured, and our instructions were to make sure they all had dancing partners

throughout the evening, 'do not let your own features display horror or undue sympathy, in fact act as though they were the most handsome of heroes'. They were heroes all right, but one really had to concentrate to manage the other part. I expect that with the development of plastic surgery their lot probably improved with time.

We always tried to look as glamorous as possible on these occasions, curtain after curtain, being sacrificed for evening skirts, not to mention bedspreads. The old Singer sewing machine was never busier. Occasionally one was given part of a parachute, what joy to have that silk to make under-clothes. We passed outgrown children's clothes round and round the circle. Bad luck if you were the youngest in a large family of cousins. You never got a chance for a new dress. There was a brand of oiled wool available without clothing coupons intended for sea boot stockings. I think it was sheep's wool, unwashed, but warm.

After only five months, in September 1940, we received the invasion alert. Although the general orders from the Government were to 'stay put', the members of the establishment were ordered to leave. My husband sowed a 'final' row of carrots—fortunately the threat of invasion passed. More than a year later (see next chapter), in the spring of 1942, the establishment was forced to leave the Dorset coast. I remained in Swanage—a son Bryan had now been added to our family—and continued with my school. Eventually, after a few months, since flats were made available in Great Malvern, I decided to join my husband, but in spite of the increased bombing of Swanage for me it was hard to leave—like the tearing out of roots.

Joyce Lovell at a war-time picnic.

Malvern, in Worcestershire, although comparatively secure from air raids, had no welcome at all for us, though eventually I started another school there and we made local friends in that way. The mother of one of my pupils had very generously passed on some of his toys and books for the use of all the others. I always tried to make a birthday cake for each child as the occasion arose. This time I seemed to have fewer rations than ever before. No sugar, no eggs (not even dried eggs) no fat—what a dreadful cake it was, how could it be otherwise? In desperation I removed the rubbery centre of the cake with a circular pastry cutter, and in the hole which was left I placed a small glass jar. I filled this with buttercups and daisies. Next day I was surprised by a message brought from his mother, 'James has never stopped talking about that wonderful birthday cake you made for him. How did you achieve it, and may I dare to ask for the recipe?' I must have looked utterly blank. 'Or is it a very precious *secret* recipe belonging to your family?' I hope she understood when I said yes it was.

Their Majesties King George VI and Queen Elizabeth visited TRE on 14 July 1944. The Queen is in conversation with W E Burcham and the King with F C Williams (back view) and P I Dee. A P Rowe is to the right of Dee.

Towards the very end of the war, amidst conditions of the closest secrecy, the establishment in Great Malvern was to receive a visit from Their Majesties The King and Queen to see for themselves some of the work that had been carried out there. I decided to take the school for a walk that July morning. Past which guarded gate? At what hour? I suppose there were about 15 children, aged five to six years. I had warned them to watch for that large car carefully, but not to expect any crowns. When we returned home I asked the children to draw a picture of what they remembered. The royal group had been soberly dressed but pinned to the lapel of Her Majesty's coat was a small and brilliantly sparkling brooch. In one of the drawings the whole page was dominated by the beauty of those shining diamonds. All the human figures were indicated without detail. I have the child's drawing still, a reminder of the shining courage which made radiant the whole picture.

Chapter 12

The Last Days in Swanage and the Food Queues of Great Malvern

In the last days of May 1942, TRE evacuated its various premises around Swanage and Hurn airport in considerable confusion and as Rowe[1] wrote:

> [in Swanage] there were no great living and housing problems. Altogether, we were as comfortable as we could expect to be in the middle of a war. Then came the bombshell, unheralded by rumour. There were, we were told, seventeen train-loads of German parachute troops on the other side of the English Channel preparing to attack T.R.E. The Prime Minister, we were told, had said that we must move away from the south coast before the next full moon. A whole regiment of infantry arrived to protect us; they blocked the road approaches to our key points, they encircled us with barbed wire, they made preparations to put demolition charges in our more secret equipments and, in the execution of their lawful duties, they made our lives a misery. Nor was this all. Our Home Guard, of considerable strength, was expected to be on duty all night and work all day. My own time was spent less in dealing with the work of T.R.E. than in discussions on whether we should die to the last scientist or run and, if the latter, where.

Although Rowe states that the move was 'unheralded by rumour' that is not correct. Six weeks earlier my diary reads 'April 14th. Well founded rumours about a TRE move. Hodlin [an administrative assistant] told me it was a new Cabinet decision to get us out of the way of the expected German raids. Marlborough College being suggested. Said to be expected in about 3 weeks', and even earlier on 4 April:

> TRE expecting a raid, part of a German airborne division and landing troops are on the Cherbourg peninsular. Enormous defence preparations and heaps of soldiers around TRE. Home Guard also on duty. Thursday night [April 2nd] was suitable and everyone thought it had started with the alert but last night and tonight the weather is bad.

119

My prosaic diary entries continued in this vein divided between the weather, the games of tennis, the progress with the Halifax and the new interest the Germans appeared to be taking in TRE, '... [April 13th] at 7.15 3 ME109's machine gunned and bombed Worth without doing much damage. Repeated Wed. lunchtime when they killed 2 service people and injured 2 in the workshops with cannon'.

The origin of this German interest in TRE manifested by the air attacks on Worth and the assembly of an invading force on the Cherbourg peninsula was the British raid on their radar installation at Bruneval during the night of February 27–28. W B Lewis (Rowe's deputy) had been consulted about the desirability of securing details of a German radar installation on the French coast.[2,3] He had supported the project to the extent of agreeing that a radar expert from TRE should be included in the raiding party. On 1 March I noticed that 'Priest [one of my tennis partners] disappeared on a "secret mission" one day last week—two nights ago we raided a radio station in N. France, Navy cum Army cum parachutists. He was in the Navy'.

According to R V Jones[4] at the end of the war when German appreciations of this Bruneval raid were discovered 'they surmised that a new phase in the Hochfrequenzkrieg (high frequency war) had begun, and that henceforward aggression of all forms could be expected against radar'. The question as to whether the Germans planned to make a raid on Swanage and TRE as part of that new aggression does not appear to have been positively resolved. The opinion of R V Jones[5] is that there was evidence that 'a German paratroop unit had moved up somewhere behind Cherbourg, and it was just a possibility that they might have Swanage in mind, but we had no positive evidence.'

In any event at that time the sporadic air attacks on Swanage and TRE intensified. 'April 26th—awakened at 7.15 a.m. by cannon fire and machine gunning. 2 ME109's attacked the station [railway] and dropped bombs which knocked down Montague Purchase's shop [our grocers].' After months of relative peace the nights became disturbed. Bath was heavily bombed, the sporadic bombs around us, the roar of the Luftwaffe overhead at night, the nuisance of barbed wire and the army, the transport of our trailers of secret equipment every night beyond the Corfe Gap of the Purbeck Peninsula soon conditioned us to the move that Rowe referred to as 'a bombshell, unheralded by rumour'. In fact on 24 April—a month before the event—I noted that 'Malvern is now at the top of the priority list for moving to. All very disturbing'.

Indeed at this critical period in the early flights of H_2S in the Halifax our existence became very disturbed. On 1 May an official statement was issued that we were to move to Malvern College and by the afternoon I was on my way with twenty others to begin the scramble for suitable accommodation for our groups in the college. But that was not the only problem. '[May 1st].

Terrible start as we were met by the billeting officer and just billeted. Ch. Smith and I got a billet in a 5th rate commercial hotel.' Even so I thought it was going to be 'a marvellous place to work but domestically simply awful. The only plans at present are billet and feed out'.

When Rowe learnt of the instruction to move TRE from Swanage and away from the south coast a frantic search began for suitable inland accommodation. According to Rowe,[6] after many false starts, it was Vivian Bowden (later Lord Bowden of Chesterfield) who suggested Malvern College in the belief that it was empty. In 1942 Easter day was 5 April and when Rowe visited Malvern in mid-April he discovered that it was, indeed, empty —but only because the pupils were on Easter vacation. The publication by Captain Spencer Freeman[7] in 1967 of the account of this move of TRE to Malvern casts doubt on Rowe's statement that the choice of Malvern College arose from Bowden's suggestion to him. Spencer Freeman was the head of the Emergency Service Organisation founded by Lord Beaverbrook when he became Minister of Aircraft Production in 1940 with the primary purpose of restoring production in factories which had been put out of action by enemy bombing. According to Spencer Freeman, after the Bruneval raid:

> In April 1942 information was obtained that a heavy scale bombing and parachute raid on the main Units of the radar research establishment at Swanage was imminent. Finally, the situation became intolerable and in April 1942, the Chiefs of Staff Committee directed the removal of both units [that is TRE and ADRDE from Christchurch] before the next full moon, say 30 days. This assignment fell to me personally.

Spencer Freeman states that he was given a list of schools and partly completed hospitals by E de Norman of the Ministry of Works and Buildings but none was suitable. Then, caught in a heavy air raid in Bristol:

> I was much frustrated by the accommodation I had seen so far. I remembered that Oscar Browne, who was the director at MAP representing TRE and with whom I had spoken before I left London, had mentioned Malvern among other places ... I decided to chance Malvern.

Malvern College had already been evacuated two and a half years earlier to make room for the Admiralty in case the evacuation of London became necessary. Subsequently the College had returned to Malvern and the story of the second evacuation of the College (to Harrow) has been described by R Blumenau.[8] The Headmaster H C A Gaunt was disturbed by the visit of government inspectors on 25 April who after touring the buildings left without giving any reason for their visit. He immediately visited the Ministry of Works and was informed that:

> owing to an unexpected development in the course of the War a certain vital Government Department had been compelled to move their establishment

elsewhere immediately; that this was a War Cabinet decision and that Malvern College had been selected as the most suitable premises in the country. The Ministry of Works and the Board of Education were making the most strenuous efforts to stop this second requisition. Mr. Butler himself (then President of the Board of Education) would be pleading the cause of Malvern at a special committee which the Cabinet had set up under Sir John Anderson, the Lord President of the Council (whose son, incidentally, had been at Malvern in the 1920s). The Committee would be meeting in two days' time. When it did, Mr. Butler's advocacy was unavailing.

The trouble with Malvern as seen from TRE was not the accommodation for the laboratories but accommodation for more than a thousand staff, many of whom were living with their families in reasonable comfort in Swanage. During May a great deal of unrest developed in default of any assurance that there would be living accommodation. Rowe became concerned and after a nugatory visit to Malvern he spent a good deal of a Sunday 17 May with Dee trying to persuade me to visit Malvern 'in an effort to do something about the living'. I have no record of why he thought that I might succeed where he failed. I clearly did not think so and after further arguments he phoned CTE (the Controller of Telecommunications Equipment) who agreed to receive a deputation of Rowe, Dee, Cockburn and myself imme-diately. On the morning of 19 May (less than a week before the date assigned for the move) we must have convinced CTE that unless most urgent atten-tion was given to the problem of housing and feeding the TRE staff it would be useless to attempt the move to Malvern. He promised to intervene with the Minister (of Aircraft Production) and did so to the extent that the Malvern Winter Gardens were commandeered and converted to a huge canteen to feed the thousand TRE staff. Spencer Freeman was in charge of the operation and he has described[9] the task of the conversion of the College and the search for billets. Contrary to the impression given by Spencer Freeman a chaotic and unpleasant atmosphere awaited us.

We spent day after day at Leeson filling seemingly endless crates with our equipment: 'Whit Monday May 25, very cold and very wet. We loaded in the morning: we had finished with Leeson. Afternoon in a barren office of Lewis' discussing H [a precision short range navigational aid]'. The sporadic bombing at night added to the disarray and my last act before I left Swanage on the morning of 26 May was to erect a bed for the family under the well of the stairs. At five I arrived at Malvern College which:

> presented an unbelievable phantasmagoric spectacle. There were hundreds of Pickfords vans. Cranes, crates, straw, lorries, mud, rain and people every-where. The billeting office was unapproachable ... so I tramped to the science block [the Preston Science School where we were assigned the top floor]. This could have been much worse. One telephone in the whole building; at least some power of sorts.

Blumenau[8,10] has described that scene from the aspect of the College.

> builders were swarming all over the College grounds which, with astonishing speed, began to sprout new buildings on every available open space except the Senior Turf. At the same time the houses were transformed. All but three were converted into laboratories and workshops. Wooden partitions disappeared—that was the end of the cubicles in the dormitories of most of the Houses—,hundreds of miles of new wiring were installed, the entire heating systems of the Houses and of the Main Buildings were scrapped and more powerful ones installed throughout, complete with new boilers and brick boiler-houses. Girders were erected in many places to strengthen floors, A.R.P. installations and emergency water tanks appeared, blastproof walls were built, three electric power stations were installed. Within six weeks a huge steel-girdered workshop with 14″ brick walls had been roofed and equipped, and a large canteen capable of feeding 1,500 people at least at one sitting had been completed, the whole of the grass space between the Science Schools and No. 3 was bristling with huts, and in many parts of the grounds strange buildings of a special design rose up.

On that first evening in Malvern, 'At last I got my billeting card—a dark cavernous, depressing house with hostile people'. Indeed our major problem was the hostility of the people of Malvern. The war had made little impact on the town and our arrival was a shock to the inhabitants who, for security reasons, could be given no idea of who we were or of what we were doing. The few hotels were full of aged ladies with their attendants or nurses who regarded us with distaste. As Rowe[11] found:

> Potential billetors fell ill with alarming regularity and the number of destitute aunts who were being given permanent homes in a few days' time passed all bounds of reason. Some gave shelter on the understanding that billetees were in by ten o'clock at night while others gave it on the understanding that they stayed out, somewhere, until the same hour.

However, whereas some of the billetors faced their obligation by placing beds for the billetees in garages or on the drive there were striking contrasts in the efforts to make it possible for us to get to work again. Although I found that it was 'all too wet and miserable', nevertheless the food at the Winter Gardens was 'an incredible, organization staffed by WVS and serving 1000 people of all sorts. The worst part is the queuing, but the psychology is good; they gave good food, and plenty of cheese and butter. After this we felt better. Hensby and I found a pub, drank a pint with Downing and then at 10 climbed up to the Worcester Beacon 1395 ft; and decided it was a marvellous place to experiment on'.

Alas, my two colleagues of that first night in Malvern were soon to be killed under tragic circumstances. The death of Downing in December of that year I have already described in Chapter 7. Less than two weeks later the crash of our Halifax bomber was to kill Hensby.

The TRE food queue at the Winter Gardens Pavilion, Great Malvern,
early June 1942.

From left to right: P I Dee (in raincoat), A L Hodgkin, and unidentified
(in raincoat, backview) talking to W E Burcham in the TRE food queue
at the Winter Gardens Pavilion, Great Malvern, early June 1942
(enlarged view of the TRE food queue above).

References

1 Rowe A P 1948 *One Story of Radar* (Cambridge: Cambridge University Press) pp 128–129

2 For a detailed account of this raid see Jones R V 1978 *Most Secret War* (London: Hamish Hamilton)

3 Churchill Winston S 1951 *The Second World War* vol IV *The Hinge of Fate* (London: Cassell) pp 248–249.

4 Jones R V 1978 *Most Secret War* (London: Hamish Hamilton) p 246

5 Personal communication Jones R V/Lovell 2 December 1988, deposited with the Lovell papers (on W B Lewis) in the John Rylands University Library of Manchester.

6 Rowe A P 1948 *One Story of Radar* (Cambridge: Cambridge University Press) p 131

7 Spencer Freeman 1967 *Production under Fire* (Dublin: C J Fallon) p 41 *et seq*

8 Blumenau Ralph 1965 *A history of Malvern College 1865–1965* (London: Macmillan)

9 See reference 7

10 See reference 8, p 141

11 Rowe A P 1948 *One Story of Radar* (Cambridge: Cambridge University Press) p 136

Chapter 13

The Crash of the Halifax Bomber

No wartime member of TRE will recall the last days of May 1942 with much pleasure. The agreeable ambience of Swanage had suddenly vanished and now 'the whole of TRE [was] queueing for food, talking and complaining about billeting and the hardships thereof'. However, for me, this local chaos soon became a trivial matter engulfed in a terrible disaster.

After three days we had our H_2S equipment working on the bench on the top floor of the Preston Science School and our two Halifax bombers had arrived at Defford—the aerodrome allocated for the use of TRE, a pleasant seven mile drive east of Malvern. The experimental magnetron H_2S was fitted in V9977 and the EMI engineers were testing their prototype equipment using the klystron in R9490. The major problems remained as they were in the last flights before we left Swanage and Hurn airport—that is ranges of only a few miles on towns and disturbing gaps and fades in the CRT picture implying that the polar diagram of the truncated scanners was awry. High priority had been given to producing a sealed-off klystron of greater power than those so far available. Although the klystron in the EMI prototype was capable of producing a peak power of 5 to 10 kilowatts none of us believed that the Air Staff directive of ranges of 15 miles from 15 000 feet could be achieved with that system. The experimental equipment in V9977 with the AI Mark VII magnetron head was a different matter. Hensby and O'Kane had obtained satisfactory responses from towns with a similar transmitter in the Blenheim at ranges of 25–30 miles and there was no reason why the magnetron equipment in V9977 should not meet the Air Staff requirement of 15 miles. Indeed, in the early flights from Defford more careful adjustments to the system, and particularly to the feed arrangements to the scanner, had led to encouraging improvements in the presentation.

Over the weekend of 6 and 7 June we had arranged to meet Blumlein and his senior staff at Defford to continue the discussions about the details of the prototype. They were staying at an hotel in Tewkesbury and after they left Defford on the Saturday evening I flew in the Halifax with Hensby.

126

Responses from Gloucester, Cheltenham and several other towns were clearly displayed at greater ranges than I had seen before and it was natural that, on the Sunday, the EMI team decided to see for themselves how the magnetron equipment was performing.

Arrangements were made for Hensby to give a short demonstration of our equipment in V9977 to Blumlein and his two EMI associates, C O Browne and F Blythen. Hensby was accompanied by Pilot Officer C E Vincent (another member of my group) and Squadron Leader R J Sansom a Bomber Command Liaison Officer on loan to us. The Halifax left Defford shortly before 3 pm on the Sunday afternoon (7 June) and since we were in the middle of a meeting with Martin Ryle and others on the subject of test gear for the H_2S equipment, Blumlein said they only needed a short flight. After two hours only those of us waiting to continue the discussion became somewhat anxious. The RAF staff, accustomed to long bomber flights, were unconcerned and the news that the Halifax had crashed did not reach Defford until 7.35 in the evening.[1] An hour later we had the tragic news that all on board including the 5 RAF crew had been killed. The bomber had crashed in the Wye Valley and at 9 pm that evening Group Captain King (the Station Commander) drove O'Kane and myself to the wreckage of the bomber to salvage what we could of our most secret equipment. Only the magnetron was recognisable.

I was still a young man in my twenties, accustomed to reading about the slaughter in the war, but not to the firsthand acquaintance of colleagues lifeless under sheets near the charred remains of a bomber in which I had flown unconcerned the previous evening. It was after midnight before we arrived back at the deserted Defford aerodrome and it is, perhaps, hardly surprising that I believed this to be the end of the H_2S project. We had lost important members of my small group, the key EMI staff, including the genius of Blumlein, who were to manufacture the equipment, and our only working H_2S equipment which was still far from meeting the Air Staff performance criterion. However, I was still unaware of the power of the Prime Minister and Cherwell, or of the desperate needs of Bomber Command. Unknown to us the Prime Minister had been pestering the Secretary of State for Air throughout the spring of that year ... '(14 April) We are placing great hopes on our bomber offensive against Germany next winter, and we must spare no pains to justify the large proportion of the national effort directed to it. (6 May 1942) I hope that a really large order for H_2S has been placed and that nothing will be allowed to stand in the way of getting this apparatus punctually', and ironically on the day of the Halifax crash, apparently misinformed about our difficulties in making H_2S work:

Prime Minister to Secretary of State for Air   7 June 42

I have learnt with pleasure that the preliminary trials of H_2S have been extremely satisfactory. But I am deeply disturbed at the very slow rate of progress promised for its production. Three sets in August and twelve in November is not even beginning to touch the problem. We must insist on getting, at any rate, a sufficient number to light up the target by the autumn, even if we cannot get them into all the bombers, and nothing should be allowed to stand in the way of this.

I propose to hold a meeting to discuss this next week and to see what can be done.[2]

This pressure from the Prime Minister was to have a dramatic effect on the rebirth of H_2S from the charred wreckage of the bomber that crashed on that day but, first, what caused the crash of Halifax V9977?

The Last Moments of Halifax V9977

It is curious that after the account in my diary, written the day after the crash, there is no reference to any investigation or discussion about the cause of the accident. I think there were probably two reasons for this. First, the pressure to re-create H_2S as a working device became relentless. Second, I suspect that Dee, the other senior members of TRE and the RAF staff at Defford were understandably anxious that the topic should not be discussed with those of us who were left to carry on with the task. In any event more than 40 years later W H Sleigh consulted me about an investigation he was making into the crash of the Halifax. Sleigh was an aeronautical engineer on the staff of RSRE (the successor to TRE) and when he retired in 1984 he decided to make a detailed investigation of the events that led to the crash of the Halifax 42 years earlier. In September 1985 Sleigh kindly sent me a copy of his remarkable study of that tragic event and the following summarises his conclusions.[3]

The Halifax took off from Defford at about 1450 hrs (7 June) and crashed near Welsh Bicknor, Herefordshire at 1615 hours. Sleigh found that the information about the cause of the crash held by the Ministry of Defence Air Historical Branch 5(RAF) was that the Halifax V9977:

> Caught fire in air 2/3000′ lost height into ground. Failure of starboard outer engine in flight which subsequently caught fire. The aircraft crashed into the ground from a height of approximately 500 feet. The engine failure and resulting fire, due to fracture in fatigue of an inlet valve stem, C of S. Caught fire in an attempt to restart engine. Fire extinguishers failed to operate and evidence suggests bottles not filled. OC: — Possibly attempted to restart engine supply power for special equipment to enable experiment to be continued. Fire extinguisher bottles probably left makers empty; steps to ensure inspection on aircraft and periodically? (We are not certain of the last word)

The Halifax could have flown safely on three engines and returned to

Defford and this led to the presumption in the report that the pilot attempted to re-start the outer starboard engine because the alternator supply to the H₂S equipment was on that engine. Sleigh's comment on this official report was that:

> The only tangible information of technical worth in the above record is reference to an inlet valve fatigue failure which is consistent with an induction system flash-back and associated major fire.

The reason for the failure, and the failure of the crew to control the fire and escape by parachute, was the object of the investigation made by Sleigh in 1984–85. He was able to locate a number of key witnesses: Mr Onslow Kirby, a farmer who as a young man had witnessed the final moments of the Halifax as it crashed onto his farmland; Mr Alex Harvey-Bailey, a former senior engineer of Rolls-Royce, the manufacturer of the Merlin engines fitted on the Halifax; and other Rolls-Royce engineers who were involved in the manufacture and test-flying of the engines during the war.

Onslow Kirby was working on the farm only 300 yards from the impact point of the Halifax. When he first saw the Halifax the starboard wing was on fire. At about 350 feet above the River Wye the starboard wing broke off, the bomber rolled over on its back and crashed in flames. The Halifax had flown over the Severn Valley to demonstrate the H₂S and Sleigh concludes that it was returning to Defford over Cardiff and Newport, and that between Newport and Chepstow the outer starboard engine suffered a flash-back. In an analysis of the ensuing rapid developments in the aircraft, Sleigh concludes that the pilot was attempting to make an emergency wheels-up crash-landing but that the fire severed the starboard wing while the Halifax was still 350 feet above ground.

Since the pilot must have known that the chances were poor of the 11 men on board surviving such a crash-landing with the plane on fire and a further 1600 gallons of petrol in the fuel tanks, Sleigh queries why no attempt was made to escape by parachute. The last brief radio message received from the Halifax was that a fire had originated when the aircraft was at a height of 2000–3000 feet. This was a sufficient height for parachute escape, and in any event the remaining three engines would have enabled additional height to be gained to facilitate such an escape. Sleigh concludes that the pilot had failed to ensure that each man carried a parachute and that he had 'decided they would all remain together for a chanced forced landing and their tragic loss must reflect on their priorities being other than personal interests'.

Sleigh's investigation with the Rolls-Royce engineers who worked on the Merlin engines led to the conclusion that it would have been impossible for the pilot to have attempted a restart of the outer starboard engine and that the catastrophic sequence of events made the question of fire extinguishers irrelevant. He requested that the existing historical record should be amended with the following correct information:

> Fatigue failure of an inlet valve stem caused by human error to ensure positive locking of that valve's tappet during a Squadron servicing check.
>
> Loss of valve allowed combustion gases to ignite the charge mixture in the engine's induction and carburation system causing a major fire which in turn crossed the power plant's fire bulkhead and ignited the fuel in the mainplane fuel tanks.
>
> The ensuing major fire caused the severance of the starboard mainplane at approximately 350 feet above ground level. The aircraft rolled to the right crashing inverted and exploded in a ball of fire killing all on board.

From the evidence in the notebooks of the Engineering Inspector at Hurn and Defford, and the Rolls-Royce records of the starboard outer engine (Merlin XX 198341), it is known that the tappets had been checked ten hours (of engine run) before the crash and it is presumed that the tappet nut was left slack at this inspection which took place sometime in the two week period in May before the Halifax left Hurn for Defford. The re-setting of the valve tappets on the Merlin engine required the adjustment and locking of 48 nuts, that is 192 per aircraft. The failure of an Air Force engine fitter to ensure the security of one of these small nuts on one of the inlet valves after re-adjustment of clearances initiated the train of events which led to these tragic deaths and to a major upheaval in our affairs.

References

1   Today, when there is precise knowledge of aircraft movements, this may seem remarkable. I am indebted to Dr O'Kane for referring me to the following detail timetable of the afternoon of 7 June 1942, as recorded in his personal ms diary.

After referring to the meeting at Defford which included Dee, Curran, O'Kane, Lovell, the EMI team and S/L Sansom who had recently arrived from Bomber Command, O'Kane's diary reads:

> Blumlein, Browne, Blythen and Sansom went up with Hensby in V9977. Airborne at 14.51 hours. As they were not down —
> 1700 Lovell rang control. Not landed.
> 1730 Rang Control. No news and no contact. No interest.
> 1745 Rang Control. No news. Rang A flight officers mess, control: Nobody about.
> 1800 Rang G/C King [the CO of Defford]
> 1810 W/C Horner [an officer on King's staff] rang up and wanted to know what was up. Told him.
> 1900 Nothing done.
> 1930 F/Lt. Reynolds [another officer on King's staff] had just left Control having informed Group.
> 1945 First news of crash.

2100 Left for Welsh Bicknor

2200 Finding the less important cargo [that is the magnetron]

2 Churchill Winston S 1951 *The Second World War* vol IV *The Hinge of Fate* (London: Cassell) pp 251–253, for this and the previous extracts

3 Annex to Aircraft for airborne radar development, by W H Sleigh. Unclassified RSRE (Malvern) document

Chapter 14

The Meeting with the Prime Minister

A week before the crash of the Halifax—during the night of 30–31 May—a thousand RAF bombers had attacked Cologne. 40 were lost. Two nights later a similar force attacked Essen and 31 bombers were lost. Within the scale of this desperate war the loss of our Halifax was statistically insignificant, but for us the loss of our colleagues and equipment was so overwhelming that I think the development would have ceased had it not been for the intervention of the Prime Minister.

On 7 June before he learnt of the loss of the Halifax the Prime Minister had sent the Secretary of State for Air the minute quoted in Chapter 13. In that minute he announced his intention to hold a meeting 'next week' for the purpose of speeding up the production of H_2S. The Halifax crash did not deflect him from this intention but the retreat of our armies in Egypt did so. He had intended to hold the H_2S meeting at 11 am on 17 June and I had gone to London with Dee in readiness. Late on the evening of the sixteenth we were told that the meeting was postponed and two days later the reason became public knowledge. Churchill had crossed the Atlantic to discuss urgent strategic issues with President Roosevelt. It was at this meeting that agreement was reached to pursue the work on the atomic bomb in the USA pooling all existing UK resources and knowledge on the subject.

During their meeting on 21 June a telegram was handed to Roosevelt:

> He passed it to me without a word. It said, 'Tobruk has surrendered, with twenty-five thousand men taken prisoners'. This was so surprising that I could not believe it, I therefore asked Ismay to inquire of London by telephone. In a few minutes he brought the following message:
> 'Tobruk has fallen, and situation deteriorated so much that there is a possibility of heavy air attack on Alexandria in near future'.
> This was one of the heaviest blows I can recall during the war. Not only were its military effects grievous, but it had affected the reputation of the British armies. At Singapore eighty-five thousand men had surrendered to inferior

132

numbers of Japanese. Now in Tobruk a garrison of twenty-five thousand (actually thirty-three thousand) seasoned soldiers had laid down their arms to perhaps one-half of their number.[1]

Churchill asked the President for 300 Sherman tanks to be shipped to Egypt. They were only just coming off the production line but were despatched without the engines installed, in six ships. The ship containing the engines was sunk by a German submarine. The President replaced them immediately but it was October before this generous gesture helped in the defeat of Rommel at Alamein.

The Prime Minister left America during the evening of 25 June, the day that a motion of censure of the Government was placed in the House of Commons. The long debate, opening on 30 June was not concluded until 2 July. In the debate the Prime Minister had spoken of the 'greatest recession of our hopes since Dunkirk' and it was against this background that he held the postponed H_2S meeting on the morning following the defeat of the 'No Confidence' motion by 475 votes to 25.

The background to H_2S seemed similarly hopeless. It was nearly four weeks since we had been able to fly. Our own magnetron H_2S equipment had been destroyed in the crash and the EMI team who alone knew the details of the equipment in Halifax R9490 had perished. During the evening of 2 July, while the debate was still in progress I travelled to London with Dee and, since we seemed on the point of losing Egypt, 'we could hardly believe that he [the PM] would waste his time on H_2S'.

I am glad that I wrote an account of that meeting on the morning of 3 July immediately afterwards otherwise my memory that it was like the transformation scene of a pantomime would seem unreal. Indeed it was a transformation scene of dramatic quality and no one present left Downing Street in any doubt of the immense powers wielded by Churchill in those years.

Today if one has official business in Downing Street one has to produce passes and the policeman at the Whitehall barrier first finds out by telephone if one is expected. On the morning of 3 July 1942, there were no such safeguards. A few minutes before 11 am Dee and I walked into Downing Street and no one asked us who we were. We then walked through the door of Number 10 and there was no one to ask us why we were doing so. A number of men known to us and several whom we had never encountered were standing in an ante-room. Sir Archibald Sinclair, the Secretary of State for Air, and Colonel Llewellyn, the Minister of Aircraft Production, were in argument about how many H_2S equipments they could produce. Since no single equipment then existed and we were then doubtful whether H_2S could be made to work, this seemed to be a peculiarly unprofitable item to discuss. We were familiar enough with Cherwell, Watson-Watt and Shoenberg the head of EMI,† and we were soon to become well-acquainted with Sir Arthur Harris the C-in-C Bomber Command, the Assistant Chief of Air Staff and Air Marshal Tait. It was our first sight of Sir Robert Renwick, who was soon to become our daily contact, and there were a few others then occupying the various posts of Controller and Director in the hierarchy to which TRE was nominally responsible.

Churchill appeared wearing a boiler suit of RAF blue zipped up the front to the neck. 'Good morning gentlemen. Come in.' He seated himself, back to the fireplace, isolated on one side of the table in the Cabinet room. He had the whole of that side of the long table to himself; we faced him arrayed along the opposite side. Several he did not know. He pointed to us in turn and having been informed of our identities proceeded to announce that he must have 200 H_2S sets by 15 October. It was explained that H_2S did not exist—the only working system had crashed weeks earlier and Shoenberg said it was impossible for his firm to make 200 sets by mid-October, even if a prototype was available. The first of many unsmoked but chewed cigars was ejected over his shoulder to impact on the surrounds of the fireplace behind him.

'We don't have objections in this room. I must have 200 sets by October.'

Four times he pressed the buzzer by his right and four times the same high ranking army officer entered.

'Where are the notes I was reading before I came into this room?'

'They are on the table before you Prime Minister.'

The fourth time he believed that they were in front of him. Another buzz:

'Where is a Secretary to take minutes?'

No one came and so the Minister of Aircraft Production commenced to fulfil this function until a Major General entered,

'Prime Minister the Cabinet is waiting.'

'Tell them to wait and come here to take minutes.'

No one was offering the Prime Minister much encouragement with regard to producing 200 H_2S sets by October. Cherwell was the nearest person to him on his left. He turned,

'What does the Professor think?'

'They can be built on breadboards'

---

† In 1931 The Gramophone Company Ltd, and Columbia Gramophone Company Ltd, formed and registered a Holding Company 'Electric and Musical Industries Ltd' (EMI). Shoenberg (later Sir Isaac) was the director of research of EMI. Sir Robert McLean who was also present at the meeting with the Prime Minister had been appointed the Managing Director of EMI in April 1941. The initial 'crash programme' H_2S units were made in the EMI research department. Subsequently the main production of later Marks of H_2S was moved from the research department to the old Rudge-Whitworth factory in Dawley Road, Hayes.

announced Cherwell to our complete astonishment. The PM diverted his attention to Harris.

'What does the Air Marshal think?'

We were well-acquainted with Harris' fierce objections to the presence of unreliable gadgets in his aircraft and his meek response, 'I must have them in Stirlings', finally turned the proceedings into a level of fantasy that bore little relation to the real world outside the Cabinet room.

Now it was nearly 1 pm. The PM said he must see the Cabinet and we must decide how to produce 200 H$_2$S sets by mid-October, 'Our only means of inflicting damage on the enemy'. Supported by Cherwell's assertion that it could be done on 'breadboards' he despatched us to an adjoining room and said we were not to emerge until we had agreed how to equip two squadrons with H$_2$S by October.

The days following the PM's meeting became crowded with journeys and meetings in an attempt to organise the production of 200 sets of an equipment that had not yet worked and for which no complete bench prototype existed. After leaving Downing Street I was driven to Nash & Thompson to discuss how they could produce 200 scanners. On Saturday morning CTE† assembled the key individuals in his office. Even if a working prototype had existed, Churchill's demand for two squadrons by October was an impossible target. CTE settled on an improbable but slightly less impossible target of 200 by the end of the year. EMI, to where we journeyed on Saturday afternoon, agreed to produce 50 of these equipments and RPU (the radar production unit set up by TRE) promised the remaining 150. A target to commence fitting the equipment in the bombers in September, only two months ahead seemed a fantasy, particularly since the demand was for Stirlings and our only experience had been with the Halifax. So the next week Shorts were added to the visiting list and on the Wednesday following Churchill's meeting after again circulating between them, EMI, Nash & Thompson and the Air Ministry, I arrived back in Malvern just before midnight, according to my diary, 'feeling quite dead'.

None of these pressures, nor the arrival of a replacement Halifax W7711 at Defford on 26 June for the crashed V9977 could have made H$_2$S a reality had they not been accompanied by other circumstances. The most important of these concerned the klystron/magnetron dispute. From the foregoing it will have been evident that we were of the opinion that it would be impossible to approach the Air Staff performance target if we were forced to use the low power klystron then available. On 10 July—a week after the PM's meeting—Cherwell and Tait‡ came to Defford and at the end of that day of

† CTE are the initials in my diary, but this was Renwick (see Chapter 15) who was CCE—the Controller of Communications Equipment.

‡ Air Vice Marshal Sir Victor Tait (then Group Captain Tait) had recently been appointed Director General of Signals (DGS) in the Air Ministry. For the remainder of the war he was one of our most helpful and influential contacts in the Air Ministry.

argument I thought we had 'almost persuaded them to use the magnetron'. Indeed we had. On 15 July the Secretary of State for Air ruled that development work on the klystron for H_2S should cease and that the two H_2S squadrons should be equipped with the magnetron version—but that a decision on their operational use would depend on the war situation: 'if the Russians hold the line of the Volga'.

In fact, we had abandoned the klystron immediately after our 10 July meeting with Cherwell and Tait. The klystron unit was removed from the Halifax R9490 and on 12 July the first flight was made in this bomber, now using a magnetron as a transmitter, a spark gap modulator, and the remaining EMI prototype units. Efforts to make a satisfactory device to destroy the magnetron in the event of the loss of a H_2S bomber over enemy territory were intensified at RAE. Eventually we learnt of the results of these tests. In the most successful trial the magnetron unit was installed in a crashed German JU88. A 10 foot hole was blown in the side of the JU88 but the appropriate expert at RAE was able to reconstruct the essential dimensions of the magnetron from its fragments in a very short time. After that the destructive device was confined to a small detonator so placed as to prohibit the immediate use of the valve should it fall into enemy hands. Curran[2] who was in charge of these trials at RAE has given his own account of these attempts to destroy the magnetron.

Now, with a Halifax airborne with the magnetron and EMI engineered units, we had a system that we believed in and the problem was now to make it work and produce enough of the units for installation in the Bomber Command aircraft before the end of the year.

References

1 Churchill Winston S 1951 *The Second World War* vol IV *The Hinge of Fate* (London: Cassell) p 343
2 Curran Samuel 1988 *Issues in Science and Education Recollections and Reflections* (Parthenon Publishing Group) p 18

Chapter 15

Bennett and Renwick

I very much doubt that if we had faced this task alone we would have succeeded in meeting the target of having H_2S in the bomber squadrons by the
end of 1942. A complex of scientific, technological, political, industrial and
operational issues were involved. That we were able to work ourselves
through this maze depended substantially on two men who were injected
into the task at that critical moment—one, a Bomber Command pilot, and
the other a wealthy industrialist.

DCT (Don) Bennett

I first met Don Bennett on 5 July, only two days after the PM's meeting.
He was then a Group Captain, just promoted from Wing Commander after
an epic escape from Norway. On 27 April 1942 Bennett had led Halifax
bombers in the earliest attack on the German battleship, *Tirpitz*, which was
moored under a precipice in an inlet of Trondheim Fjord, Norway. His aircraft was hit by anti-aircraft fire but with his second pilot he escaped by
parachute and landed in enemy-occupied Norway. After an arduous journey
across snow-covered mountains and across icy streams they reached Swedish
territory and eventually made their way back to Britain. Bennett was not
only a very brave man, he was an inexhaustible and dynamic Australian. In
1938 he had set a new record for an east to west crossing of the Atlantic and
after the outbreak of war he led the first ferry flight of American and
Canadian aircraft to the UK. In their official history of the strategic bombing
campaign Webster and Frankland concluded that Bennett was 'perhaps the
greatest flying expert in Bomber Command'.[1]

When Bennett arrived at Defford he had just been placed in a very strong
position—as Commander of the newly created Pathfinder Force of Bomber
Command. The creation of this Pathfinder Force had been a contentious
issue. For some time the feeling had been growing in the Air Ministry that
Bomber Command needed a specialised 'target-finding force' and that it
should be created by forming a number of squadrons of the best crews from
the existing Bomber Command groups and placing them under the

command of one man. Harris was strongly opposed to the concept, believing that it would foster jealousy and make the crews of the main force feel that they were inferior. Harris was well aware of the problem of accurate bombing but believed that the solution would be to create a specialised target finding force within each group. He was overruled by Air Marshal Portal, the Chief of Air Staff, and in the third week of June 1942 Harris agreed to the general concept of the special target-finding force to be known as the Pathfinder Force.[2]

Air Vice Marshal D C T Bennett CBE, DSO, Air Officer Commanding 8 Group RAF (the Pathfinder Force). (Copyright Universal Pictorial Press Agency.)

Although the concept of the Pathfinder Force was imposed upon Harris, Don Bennett was his nominee as the Commander of the Force. Bennett had served under Harris before the war, moreover he held nearly every licence available in civil aviation—including the certificates of First Navigator, Radio Operator and Engineer. The coincidence of the creation of the Pathfinder Force and the appointment of Bennett as its Commander, with the critical evolutionary stage of H_2S transpired to be a matter of great significance. Bennett arrived at the HQ of Bomber Command to assume his new appointment on 3 July and was immediately sent to Defford to participate in the airborne trials of H_2S. He installed himself in the mess and demanded from the Officer Commanding at Defford that any servicing of the Halifax should be carried out to suit his flying programme. My diary note conveys the atmosphere: 'Bennett lived at Defford and on Sunday, Monday, Tuesday and Wednesday organised the Halifax to fly at 6 am and often landed at 11.30 pm. By Tuesday night I was almost hysterical ...' The only H_2S equipment was the klystron version in Halifax R9490 and when we changed this to the magnetron, 'by Wednesday Bennett managed to get a fairly successful magnetron flight'. His experiences with the klystron versus magnetron equipment had been a vital factor in persuading Cherwell and Tait during their visit on 10 July that the klystron should be abandoned.

Bennett's fervent belief in H_2S became a critical factor in the subsequent stages of its operational use. For the next three years after his first visit to Defford in early July 1942 he became almost a part of our daily lives, either there, at his HQ in Huntingdon, or on the operational airfields of 8 Group —the Pathfinder Force.

Robert Renwick

On 6 May when we were in the early stages of flying the experimental H_2S system in Halifax V9977 at Hurn airport, Churchill had minuted the Secretary of State for Air[3]: 'I hope that a really large order for H_2S has been placed and that nothing will be allowed to stand in the way of getting this apparatus punctually' and, in connection with this and other matters concerning the production of the bombers, he added:

> there are so many facets of the task which have to be completed at the proper time that it might be a good thing to appoint some one man to be responsible for taking the necessary action by the proper dates and rendering a monthly report. I have heard Sir Robert Renwick mentioned as a man of drive and business experience who has already rendered valuable service in connection with 'Gee'. Perhaps you might think he is a good man for this purpose.

When the war began Renwick was the Chairman of the London County Electric Supply Company. In the autumn of 1940 Beaverbrook, the Minister of Aircraft Production, had asked him to coordinate the production of the new four-engined bombers. He acted in this capacity with considerable success—even to the extent of arranging for experts from his London Electric Supply Company to assist with the solution of electrical problems in the Stirling bombers. In Chapter 8 reference has already been made to his appointment as Chairman of an RDF Committee in the Air Ministry responsible for the coordination of all research, development and production of radar and radio aids in aircraft. He assumed this responsibility in October 1941 shortly after the analysis of the RAF night bombing raids revealed their ineffectiveness. He was responsible for the decision to form an RDF (radar) department at Bomber Command headquarters and for the appointment of Saward to command this post. The Secretary of State for Air therefore had little difficulty in adding H_2S to Renwick's responsibilities in response to the PM's minute of 6 May 1942.

Sir Robert Renwick Bt held the joint war-time posts of Controller of Communications in the Air Ministry and Controller of Communications Equipment in the Ministry of Aircraft Production. (Photograph courtesy of Group Captain Saward.)

Although, as mentioned in Chapter 8, it seems possible that it was Renwick who stimulated the Sunday Soviet in Rowe's office on 26 October 1941 which was to lead to the development of H_2S, I have no record or recollection of encountering him until after the crash of the Halifax. Then:

> June 28th. Sir Robert Renwick and his man Sayers appeared, appointed by the Minister of A.P. [Aircraft Production] and the Sec. of State at the request of the P.M. ... most impressed by Renwick.

Frank Sayers was described to me on one occasion as 'Renwick's winger' and for the remainder of my wartime life he certainly acted in that capacity. Many of our problems with the supply and manufacture of our equipment were instantly dealt with by Sayers. If he encountered obstacles, whether it be in the supply of a bomber or a thermionic valve, he would inform Renwick. If Renwick then encountered obstructions he would telephone the PM.

With this highest priority Renwick and Sayers quickly assessed what they considered to be a realistic target and on 24 July a programme was established and a memorandum issued by TRE giving the detailed schedule for equipping 24 Halifax and 24 Stirling bombers with H_2S by 31 December. At this time in the summer of 1942 two new posts were created, that of Controller of Communications (C of C) in the Air Ministry and Controller of Communications Equipment (CCE) in the Ministry of Aircraft Production. Renwick was appointed to both and throughout the war he remained an immensely powerful civilian in these twin posts straddling the two ministries with which we were involved. The contact between Renwick and Sayers and ourselves in TRE became a daily one until the crisis of the war was over. Every week he would summon the vital 50 people concerned with H_2S to his office in the Air Ministry. Between these meetings he would telephone either Dee or myself, 'Any news any problems?' Even after nearly 50 years I find it odd to be walking along King Charles Street and passing the door that once led to the corridors of the Air Ministry and Renwick's office.

Saturday August 8 [1942] Another Renwick meeting in London yesterday, after another week of struggle'.

The 'struggle' was that we still had to make the H_2S work.

References

1 Webster Sir Charles and Frankland Noble 1961 *The Strategic Air Offensive against Germany 1939–45* (London: HMSO) vol III footnote p 299
2 For the details of the exchanges between Portal and Harris see Saward Dudley 1984 *'Bomber' Harris* (London: Cassell) ch 15
3 Churchill Winston S 1951 *The Second World War* vol IV *The Hinge of Fate* (London: Cassell) p 253

Chapter 16

Autumn 1942

The programme of work is simply crazy and my office gets more and more full of papers, drawings and heaven knows what.

Soon after the vital mid-July decision to abandon the klystron version of H_2S we had two Halifax aircraft at Defford equipped with H_2S. Both had magnetron transmitters and, with the EMI team re-organised after the June tragedy, the major problem might have been to produce the units and install them in the bombers. The difficulties associated with the use of the magnetron, the modulator, the receiver and the TR system had already been substantially surmounted since AI Mark VII was in operational use. Special control and indicator units were, of course, needed for the H_2S PPI presentation but these presented no serious difficulties and in discussions with Blumlein, Browne and Blythen, before they were killed, we had reached agreement about the design of these units.

Unfortunately, the results when flying at 10 000 feet and above in the Halifax were poor. There was a major trouble with the polar diagram of the scanner. This was manifested on the PPI as a dense cluster of ground returns beginning as a ring at a range equal to the altitude of the bomber and extending for a few miles in range. From the point of view of the navigator there was a more serious defect. The response from a town, for example, would be seen at a range of 15 to 20 miles but when the bomber homed on to the town the radar echo on the PPI would fade, disappear and then appear again. The variable manner of these responses was very far from the clear and well-defined radar response within the 10 mile range demanded by Bomber Command.

The political and operational pressures for the system had made it impossible to pursue appropriate ground tests which would have been the normal scientific procedure to find out how to produce the correct vertical polar diagram from a truncated scanner rotating in a perspex cupola under the fuselage of a bomber. The favourable Blenheim V6000 results had been obtained with a full paraboloid mounted in the nose of the aircraft and having a clear forward view. Further, the Blenheim tests had been made at an altitude of a few thousand feet. Although a single flight in a Beaufighter at 20 000 feet had shown that built up areas still gave clear responses, there had been no appreciation of the radical difference between these full paraboloid, narrow beam, clear forward-looking installations and the performance of a truncated paraboloid suffering irregular scattering from the underbelly of the bomber.

In the event we were reduced to experimenting with *ad hoc* adjustments to the feed arrangements of the scanner with the bomber airborne at altitudes up to 18 000 feet. The TR unit was mounted immediately above the scanner and this made access to the feed and matching arrangements especially difficult and the problem was not exactly eased by the cumbersome flying suit and the need for oxygen masks. After the death of Hensby, O'Kane had borne the brunt of these airborne tests but, since he was one of the very few people familiar with the total system, we had agreed in late June that he would go with the first H_2S-equipped bomber for the service trials at the Bomber Development Unit (BDU)—at that time the target was mid-August for these trials to commence. Fortunately, Saward and his staff at Bomber Command readily responded to cries for help and soon a number of eager and enthusiastic young officers were attached to my group. Flight Lieutenants D W C Ramsay; Peter Hillman, the American; James H (Scotty) Sawyer and the Canadian Joe Richards began to shoulder the burden of the autumn flying tests.

Our first target was to get an H_2S-equipped bomber to BDU for service trials. Originally in the optimism of the early months of that year we had planned to send Halifax V9977 to BDU in May and expected the service trials to be carried out during the time when we were moving from Hurn to Defford. In the event, after a discussion with Renwick at Defford, a formal letter was sent to DCD on 30 August in which we confessed our failure to fill the gaps and cure the fades and recommended that these experimental aircraft should not be sent to BDU but that the service trials should await the first prototype equipment. There is little doubt that this saved H_2S as an operational system because another month elasped before a Halifax was despatched to BDU for these trials. Then on 30 September a Halifax W7808 with the test installation of EMI production apparatus and a somewhat improved picture on the PPI was despatched to BDU. With anxiety we awaited the reaction—and to our relief it was not altogether unsatisfactory. The BDU navigators found the serviceability 'hopelessly bad' but thought that H_2S would be 'valuable to a high extent both as a navigational aid and as an aid to locating targets'.

Opposition to H_2S

During those autumn days of 1942 our troubles were not only associated with the problem of achieving a satisfactory H_2S system. Surprisingly there was a great deal of opposition to the system. A good deal of this originated within the RAF Bomber Command from the same kind of attitudes and jealousies that led to the opposition to the creation of the special Pathfinder Force. At least in the beginning H_2S could only work within the Pathfinder Force and it was, perhaps, to be expected that this would increase the antagonism between the Bomber Command groups. For example, 5 Group led by Air Vice Marshal The Hon R A Cochrane had many skilled crews and navigators who claimed to be able to reach their targets using conventional means of navigation. Indeed, with Harris' encouragement this group had evolved its own pathfinding and target marking techniques which differed from those used by Bennett in 8 Group.[1] Cochrane had been entrusted with the task of attacking a number of special targets (for example the Schneider works at Le Creusot on 17 October 1942) and did so with great success (on 16–17 May 1943, 5 Group carried out the famous raids on the Ruhr dams). The antagonism between Cochrane's 5 Group and Bennett's 8 Group (Pathfinder Force) had this basis, and was undoubtedly the source of much of the opposition to H_2S. These jealousies were well known in RAF Bomber Command and their extent is well-illustrated by the rumour circulating at that time that Bennett and Cochrane had challenged one another to a bombing match using each other's headquarters as the target.

In TRE those of us responsible for H_2S were greatly disturbed and aggravated by these developments, particularly when we received instructions to use the experimental systems in Halifax R9490 and W7711 to make a series of special flights. Someone had informed Harris that 'the responses as seen on the PPI did not correspond to objects and areas which one would expect to see and that if an observer homed on to the most prominent return on his tube it would in many cases turn out to be other than a worthwhile objective'. Although we were ourselves most dissatisfied with the appearance of the PPI picture these criticisms were a nonsense. Nevertheless, when our time could have been much better employed in attempting to improve the system, we were forced to divert our efforts to carry out these special flights. On 27 October 1942 we issued our report containing the evidence that negated these criticisms. Figure 16.1, a photograph of the PPI taken during one of these flights in Halifax W7711 flying at 10 000 feet, is typical of the H_2S picture obtained at that time. The bomber was flying towards Gloucester and the responses from that town, Cheltenham and an aerodrome midway between them are quite clear. So is the ring of ground clutter which remained a most disturbing feature in these early flights.

The photograph reproduced in figure 16.1 was taken by the American Air Force Lieutenant (Scotty) Sawyer who had been attached to my group. Sawyer was an enthusiastic believer in H_2S but unfortunately that could not be said about other influential Americans who had been acquainted with the plans for H_2S. Indeed, the opponents of H_2S in RAF Bomber Command seized on the American criticisms to support their own arguments against

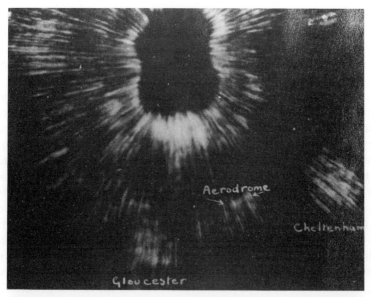

Figure 16.1 September 1942. A photograph of the CRT when the H_2S-equipped Halifax W7711 was flying at 10 000 feet. The towns of Cheltenham, Gloucester and an aerodrome can be identified, but the heavy ring of the ground returns made interpretation difficult at closer ranges. This was typical of the performance of the H_2S first sent to BDU for assessment on 30 September 1942.

the system. As Guerlac states in his American history of radar in World War II[2] the origin of this opposition is obscure.

Early in 1942 the Radiation Laboratory was informed through the British Air Commission of the results obtained in England. For reasons which have never been clearly explained, attempts to verify the British results with 10-cm American ASV equipment were not successful. Largely because of this, the American visitors in the summer of 1942 expressed rather forcibly their skepticism concerning the H_2S development. There were also strong currents of skepticism in Great Britain—some doubts even at TRE—and many difficulties in the course of development. The opponents of H_2S equipment, even at this late date, nearly outnumbered its supporters.

There were two senior TRE scientists working at the MIT Radiation Laboratory at that time. As described in the Preface and Chapter 3, E G Bowen had gone to America with the Tizard mission in August 1940 and he remained there. In Chapter 5 I have also described the association of D M Robinson with the early centimetre applications. In the spring of 1941 TRE received an instruction to send a centimetre wave expert to the MIT Radiation Laboratory in order to supervise the fitting of lease–lend aircraft with

centimetre ASV. Robinson volunteered for this task and in mid-July 1941 he
left TRE for America and by March 1942 he had the prototype centimetre
ASV airborne in a B24 Liberator. This is the system referred to by Guerlac.
Bowen's initial interest in America concerned the production of an American
version of the centimetre air interception (AI) system. He has described[3] the
impact of the magnetron in the USA and the first flights with an airborne
centimetre radar on 10 March 1942. Robinson was certainly not opposed to
H_2S and as will be seen later he returned for a time in 1944 to TRE as a temporary replacement for Dee as Superintendent of the centimetre applications, including H_2S. Nevertheless, at the critical time in 1942 neither he
nor Bowen were able to satisfy themselves from those experiences in
America that the use of centimetre equipment would provide a workable
form of H_2S.

At the height of my own worries about the poor quality of the returns from
built-up areas on the cathode-ray tube, Dee received a letter from Robinson
in which he stated that 'H_2S did not work in America'. He had flown with
a centimetre system and, although he had carefully manipulated the controls, had failed to detect any distinction between towns and ground. This
cryptic announcement was followed on 5 and 6 July 1942 by a visit to TRE
of I I Rabi (at that time Head of the Research Division and Associate
Director of the MIT Radiation Laboratory) and E M Purcell, who, in a
somewhat unpleasant meeting in Dee's office, announced that in America
they had concluded that our H_2S device was unscientific and unworkable
and that if we persisted with our plans the only result would be that the
Germans would obtain the secret of the magnetron. This depressing visit
added to our own immediate problems because the American view was
seized upon by the British opponents of H_2S.

As will be related H_2S was first used operationally over Germany with
great success at the end of January 1943. The next day a letter arrived in
Dee's office from Robinson in America suggesting that 'the time was now
ripe for settling the controversy as to whether this device will or will not
work by doing some scientific measurements'. The remarkable twist to this
story was the subsequent turnabout of the American view when our own
H_2S began successful operations.

Nearly 50 years later I attempted to elucidate the nature of this initial
American attitude to H_2S. Robinson expressed the opinion that there was no
'concerted American or R.L. [Radiation Laboratory] opposition to H_2S at
any time'[4] and indeed, as will be seen, he soon became an ardent supporter
of the eventual American system.

Whereas Robinson was expressing the reasonable doubts of a scientist on
the basis of tests carried out in an atmosphere far-removed from the political
and military pressures to which H_2S was subject in the UK, the case of Rabi
and Purcell appears in a different light. In his later years I had occasion to
seek enlightenment in conversation with Rabi but perhaps, understandably,

he did not remember the occasion or the reason for his visit to us in July 1942 that caused us so much anxiety. My correspondence[5] with his biographer John S Rigden was more enlightening. He found that neither Rabi nor Purcell had recollected the visit to TRE and the episode in Dee's office recorded in my diary but that:

> Rabi and Purcell do recall the concern shared by both British and American scientists with the use of radar in bombing missions over the Continent for fear the magnetron would fall into the hands of German physicists. There might have been heated discussions as a consequence of this concern which might have been misunderstood as opposition to the H_2S itself.

Lee Du Bridge (the director of the MIT Radiation Laboratory at that time) responded to Rigden's enquiry that:

> ... Both US and British workers had considerable trouble with the 10-cm H_2S in early bombing trials, leading many on both sides of the Atlantic to doubt the usefulness of such equipment. Rabi may have been in England at just this time, and shared these doubts. However, things moved fast on both sides of the Atlantic and the H_2S and H_2X became rapidly more successful and appreciated.

Whereas the above places the 1942 reaction of Robinson, and of Rabi and Purcell, in perspective Bowen's attitude at that time is far more difficult to understand. In 1938–39 the use of his 1½ metre airborne system for navigation and target detection figures prominently in his list of applications for airborne radar. His book *Radar Days* published in 1987 contains a detailed description of the centimetre developments in the USA following the 1940 Tizard mission, but there was no reference to the H_2S application. Following the publication of his book I corresponded[6] with Bowen (then in Australia) about the failure to 'make H_2S work' in the USA in 1942. After conceding that 'coastal targets were easy to discriminate with 10 cm radar', he continued:

> Compared with these, cities in inland USA were an entirely different proposition. For example, two townships in Connecticut—Springfield and Hartford—both of which were important defence production centres—the latter producing most of the Pratt and Whitney aero-engines in the USA at that time, were quite invisible on an H_2S radar ... in tests made in the Boston area, quite substantial townships like Worcester and Manchester simply did not stand out as targets. However, the biggest failures in the Eastern part of the US was Pittsburgh and, to a lesser extent, the Schenectady–Albany complex. These were and still remain among the most important metal and electronic production centres in the Eastern half of the USA. Neither could be seen on a 10 cms radar.

In this correspondence Bowen ascribes the failure to the nature of the terrain and claims that for equivalent reasons we would have failed in attempts to obtain responses in tests against 'Sheffield or the Bradford–Leeds area'. But this is incorrect and Bowen's failure to make 10 centimetre H_2S work in America in 1942 remains an enigma.

H_2S to the Pathfinder Force

It seems probable that a good deal of the negative American attitude to H_2S in 1942 arose because there was no significant understanding of the operational pressures. Indeed, it may be no coincidence that the *volte-face* in this attitude in 1943 occurred when the American 8th Air Force began operating in Europe, often in cloudy and unfavourable weather.

We had no time to make careful experimental assessments. Of these initial stages in H_2S development Churchill's attitude was that:

> If too much time were spent on experiment manufacture would be delayed and so would accuracy of bombing.[7]

Bowen and Robinson were carrying out scientific assessments of the possibility of H_2S far-removed from the war-front and must have been quite unaware of the most urgent operational needs. If I had doubts about the usefulness to the bomber crews of our poor PPI presentation they were completely swept away during one night in October 1942.

O'Kane had gone with the Halifax W7808, fitted with the test installation of the EMI prototype units, to BDU for the service trials on 30 September. Two weeks later:

> [14 October 1942] The work is going well. BDU are pleased and the gaps are quite gone from the Halifax. Thursday [14 October] Saward and Dickie flew down to pick me up [from Defford] and we spent the day at BDU talking to the navigators using it.

That day was a great encouragement. The skilled navigators of BDU had found even the poor quality PPI picture invaluable when flying above thick cloud. But it was that night that made me appreciate how urgently H_2S was needed.

> Thursday night [14–15 October] was a vital experience. We flew on to Lakenheath. 10 Stirlings took off for Cologne at 7 pm and I stood around in the interrogation which went on till nearly 2.30 [am]. At the end of it as fellow after fellow confessed he's never even seen the target and that all was hellishly confusing over N.W. Germany I was feeling absolutely convinced of the vital necessity of H_2S.

The problem now was to meet the Renwick target set on 24 July of 24 Halifax and 24 Stirling bombers equipped with H_2S by 31 December (see Chapter 15). The difficulties were increased by the extraordinary demand made by Harris at the PM's July meeting that he wanted H_2S in the Stirling bombers. The superior Lancaster bombers were replacing the Stirlings and Roy Chadwick, the chief designer of the Lancaster, had to be persuaded to install the perspex cupola underneath the fuselage. Chadwick was rightly proud of the performance of the Lancaster and opposed the H_2S scanner installation even more vigorously than Handley Page had done in the case of the Halifax early in January. This was one of the many occasions when Renwick's aid at a higher level had to be enlisted and early in November I made a journey through thick fog to Ringway airport for the trial installation of H_2S in a Lancaster.

Good news continued to emanate from BDU. Their complaint was not about the usefulness of H_2S but that the serviceability of the equipment was so poor. This was primarily due to the breakdown of the pulse transformer in the TR unit and the desperate efforts to improve this device did achieve some success as the H_2S-equipped Halifax and Stirling bombers began to reach their operational aerodromes in 8 Group.

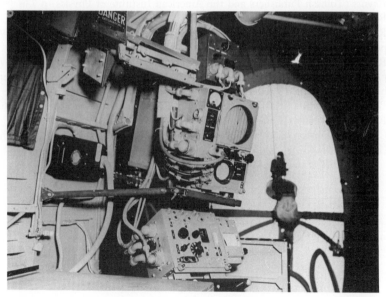

Figure 16.2 The H_2S display unit in the navigator's cabin of a Halifax.

The target of 24 Halifax and 24 Stirling H_2S-equipped bombers by the end of the year depended on the promise of EMI to supply 50 sets and of RPU to supply 150. RPU completely failed to produce any sets by 31 December, but it is a great tribute to the wartime EMI organisation that, although they had lost Blumlein and two other key members of their staff, they more than fulfilled their promise.

Churchill's demand for two squadrons of H_2S bombers by October was never remotely possible. The Renwick/Sayers realism which stimulated our schedule of 24 July for the equipment of 24 Halifax and 24 Stirling aircraft by 31 December was just a possibility—but it was only partially realised. By the last day of 1942 there were 24 H_2S aircraft in 8 Group, 12 Halifax at 35 Squadron, Graveley, and 12 Stirling at 7 Squadron, Oakington. The great battles on the Eastern front had waged throughout these autumn months. The line of the Volga was not breached (cf. p 136 Chapter 14). 'Dec. 14th. They [the Pathfinder Force] have permission to use it after Jan. 1st. This will be terribly exciting.'

References

1 For a description of this 'master bomber' technique see Saward Dudley 1984 *'Bomber' Harris* (London: Cassell) pp 238–239
2 Guerlac Henry E 1987 *Radar in World War II* vol 8 in *The History of Modern Physics* (Tomash Publishers and the American Institute of Physics) p 736
3 Bowen E G 1987 *Radar Days* (Bristol: Adam Hilger) ch 10 and 11
4 Robinson to Lovell 1 November 1990 (correspondence deposited with the Lovell papers in the John Rylands University Library of Manchester)
5 Rigden/Lovell correspondence June–September 1987 deposited with the Lovell papers in the John Rylands University Library of Manchester. Rigden's biography of Rabi is Rigden John S 1987 *Rabi* (New York: Basic Books)
6 Bowen to Lovell 14, 20 September 1987. Lovell papers deposited in the John Rylands University Library of Manchester.
7 Churchill Winston S 1951 *The Second World War* vol IV *The Hinge of Fate* (London: Cassell) p 251

Chapter 17

January–February 1943

On 15 July 1942 the Secretary of State ruled that development work on the klystron for H_2S should cease and that the two squadrons should be equipped with the magnetron the important condition was made that they would be used over enemy territory only if the Russians held the line of the Volga. At that time I wrote in my diary:

> ... the Germans are driving back the Russians towards the Volga and things are bad there ...

and a few weeks later, with Rommel thrusting towards Cairo:

> [Aug. 8th] ... the war is pretty depressing. The Germans have driven the Russians right back to the Caucasus.

However, a few weeks later the enormous battles which were to save Stalingrad began. On the shore of the Black Sea, Novorossisk was taken on 10 September but, as described by Churchill.[1] 'Hitler's orders to seize the whole of the Black Sea littoral could not be carried out ... It was not till September 15 that after heavy fighting between the Don and the Volga, the outskirts of Stalingrad were reached. The battering-ram attacks of the next month made some progress at the cost of terrible slaughter. Nothing could overcome the Russians fighting with passionate devotion amid the ruins of their city.' By the middle of October, 'General Paulus's Sixth Army had expended its efforts at Stalingrad, and now lay exhausted with its flanks thinly protected by allies of dubious quality'.

The operational use of H_2S depended on whether Hitler's efforts to relieve Paulus's army succeeded, but, by January, '... at last they began to crack, and by January 17 the Russians were within ten miles of Stalingrad itself. Paulus threw into the fight every man who could bear arms, but it was no use. On January 22 the Russians surged forward again, until the Germans were thrown back on the outskirts of the city they had tried to take in vain. Here the remains of the once great army were pinned in an oblong only four miles deep by eight long'.[2] Eight nights later, on the eve of the capture of Paulus and his staff, H_2S was first used over enemy territory.

After his experience with the service trials of H$_2$S at BDU, O'Kane had written the preliminary operating instructions and had gone with others from TRE to the aerodromes of 8 Group at Graveley and Oakington to help with the training of the navigators and with the maintenance of the equipment. '[27 January, 1943] The Pathfinders are ready to go but are held up by unserviceability due to terrible troubles with the pulse transformers'. Two days later O'Kane 'advised me to take a weekend in the country' and on the morning of Saturday 30 January Shea (a member of Bennett's staff) phoned, 'come up straight away'. I reached the aerodrome at Wyton at 5.30 that evening:

> It had been laid on for Berlin to follow the first two daylight raids by Mosquitoes which had messed up Görings and Goebbels Jan. 30th speeches but the Met was too bad and it was switched to Hamburg. The take off was midnight, zero hour 3 a.m. and return 6.30.

That was an anxious night without sleep.

> By 6 we all felt terrible. There seemed to be a complete shambles. 40 Lancs of the main force had turned back owing to appalling weather. 4 of the 6 Graveley Halifax's were back and 3 of the Oakington Stirlings, 4 with their H$_2$S u/s. Things cheered up with the dawn however and eventually 6 of our gears got right through and everything worked perfectly. Also 100 Lancs followed up and bombed over the markers. So as things finally emerged the Pathfinders were overjoyed and Bennett was in fine form. The first do had been a success (extract from diary note dated 1 February, about the night of 30–31 January 1943).

That night I was at Graveley the base of the 35 Squadron H$_2$S Halifaxes. Saward went to 7 Squadron at Oakington where the H$_2$S Stirlings were based and of that night he wrote:

> There was no moon, and visibility conditions over the target, as expected, were such that visual identification was almost impossible. The actual report on weather by crews was summed up in my note on the first operational use of H$_2$S: 'Heavy static was reported and over the Dutch coast much cloud was encountered up to 20,000 feet with severe icing. In the target area itself there was little cloud, but ground haze and the absence of any moon made visual identification almost impossible.' ...When the first H$_2$S aircraft returned, the crews were enthusiastic. They reported that they had had no difficulty in identifying Heligoland, Zwolle, Bremen, Zuider Zee, Den Helder, East and North Frisians, Cuxhaven and Hamburg itself. The average range at which towns were identified was about twenty-three miles, the maximum being thirty-three miles and the minimum 11½ miles. Coastlines, estuaries and rivers were described as appearing on the cathode ray screen 'like a well-defined picture of a map', and the six navigators who reached the target claimed positive identification of the docks, stating that they appeared as 'fingers of bright light sticking out into the darkness of the Elbe'.[3]

In their official history Webster and Frankland[4] write of this raid:

> reports [of navigators] suggested that the target area had materialised in
> exactly the expected form ... the occasion was memorable for the introduction
> of a new radar aid [i.e. H2S] which could not be jammed and which had
> unlimited range but also the first time the target indicator bombs had been
> used in combination with radar.

Two nights later (2–3 February) 10 H2S Pathfinders marked Cologne for
the main force, 11 marked Hamburg again on 3–4 February and 8 marked
Turin on 4–5 February. For a time, at least, all seemed to be going well.
A telegram arrived on my desk:[5]

> To Dr. Lovell, TRE Great Malvern.
> From O/C Pathfinder Force, RAF Wyton.
> Heartiest congratulations from myself and the users to you and your collab-
> orators on the development of the outstanding contribution to the war effort
> which has just been brought into action.

AI Mark VIII had just shot down its first raider but our celebrations of
this progress with two of the centimetre systems in Dee's division were to
be short-lived.

Of February, I noted in the diary: 'The bombing effort has been colossal
—day and night. Our stuff has gone out about 14 times now and they are
as excited as ever with it' and, after the first few operations, Bomber
Command issued a memorandum:[6]

> H2S in its present form fully meets Air Staff requirements and has exceeded
> expectations in that towns have proved easy to identify both by shape and
> relative positions. In addition to the exceptional value of H2S for identifica-
> tion and bombing of the target, its great navigational value has been proved
> beyond all doubt. The recognition of islands, coast-lines, estuaries and lakes
> has been particularly easy. In fact the problem of accurate navigation under
> almost any weather conditions is solved by H2S when operated by a trained
> navigator.

But in the conclusions of this memorandum there was also a comment that
tempered our euphoria after these first operations.

> Some re-design of the scanner system is required to prevent gaps appearing
> at heights of 20,000 ft and above. (It should be noted that the scanner was
> originally designed for maximum operational efficiency at 15,000 ft).

We were well aware that the deficiencies in the vertical polar diagram of
the scanner became increasingly manifest as the altitude of operation
increased. The performance of these first H2S systems was reasonably good
at 15 000 feet (which was the original Air Staff specification). But even in

these initial operations the Halifax bombers had flown at 18 000 feet and the operational height of the Lancasters, which were soon to form the core of the bomber force, was at least 22 000 feet.

Evidently, the acute problem of the gaps and fades in the H_2S picture on the 10 mile bombing scale remained with us but before we made much progress towards a solution we became immersed in another quite different operational use of the units that formed the basic system of H_2S.

References

1 Churchill Winston S 1951 *The Second World War* vol IV *The Hinge of Fate* (London: Cassell) pp 524–525.
2 Churchill Winston S 1951 *The Second World War* vol IV *The Hinge of Fate* (London: Cassell) pp 637–638.
3 Saward Dudley 1984 *Bernard Lovell* (London: Robert Hale) pp 77–79
4 Webster Sir Charles and Frankland Noble 1961 *The Strategic Air Offensive against Germany 1939–1945*, vol 2 (London: HMSO) p 103
5 The original is in Lovell TRE record p 8
6 *Memorandum on operational use of H_2S* BC/S 26180/16/RDF 9 February 1943 (copy in Lovell TRE record pp 9–11)

Chapter 18

Centimetre ASV *and the U-Boats*

In Chapter 3 an account has been given of my own fringe association with the early ASV system working on a wavelength of 1½ metres. That was at St Athan during the winter of 1939–40 and it was a test flight of one of these ASV systems that led to the death of my two colleagues Ingleby and Beattie when the Hudson crashed in South Wales. The early development of the metre wave ASV system has been described by Bowen[1] and a detailed account of the subsequent developments and operational use by Coastal Command has been given by Smith, Hanbury Brown, Mould, Ward and Walker.[2]

As described in earlier chapters of this book, when the move to Worth Matravers occurred early in May 1940 I became immersed in the development of centimetre radar and it seemed most unlikely that I should have any further connection with ASV systems. However, this was not the case and I must now describe the circumstances that led to the involvement of a substantial part of my group with Coastal Command, coincident with the first operational use of H_2S over enemy territory.

The U-Boat Attacks in the North Atlantic

During 1942 the shipping losses from U-boat attacks in the North Atlantic reached 600 000 tons per month. Churchill and the Cabinet were extremely concerned, and in his account of those years Churchill states that his great fear was that Hitler would decide to stake all on the U-boat campaign. By midsummer the War Cabinet considered that the losses to the Atlantic merchant fleet 'constituted a terrible event in a very bad time' and that the U-boat attacks were then our worst evil. Every history of that period reveals the critical nature of these U-boat attacks on the ships bringing oil and essential supplies to Britain. In the official history of the war at sea Roskill[3] states that in the first seven months of 1942 the Germans had sunk 681 ships — a tonnage of 3 566 999 tons and 589 of these (over 3 million tons) had been sunk in the Atlantic and Arctic theatres. Only 3.9 per cent of the U-boats at sea had been destroyed and on the average 300 tons of shipping per U-boat per day had been sunk.

After the collapse of French resistance in 1940 the U-boats were able to work from the French ports in the Bay of Biscay. If in danger of attack, they submerged by day and surfaced at night to recharge their batteries. Hence they were not seriously hampered in their passage to and from the Bay ports, and a considerable fleet was maintained in the Atlantic. Churchill had repeatedly begged President Roosevelt for help but the aid from the United States, in the form of destroyers and long range aircraft operating over the North Atlantic, barely compensated for the ever increasing size of the U-boat fleet able to operate on the Atlantic routes relatively immune from attacks from the air.

Fortunately the 'transit area' of the Bay of Biscay was within reasonable range of aircraft based in Britain, and U-boats in transit on the surface of the Bay could be detected by the ASV 1½ metre radar, which had been operational in Coastal Command aircraft since January 1941. However, this was insufficiently accurate to facilitate successful attacks in darkness, and in any event visual identification was essential in order to make sure that the radar echo was from a U-boat. In 1942, a powerful searchlight with an azimuth spread of about 11 degrees (the idea of Wing Commander H de V Leigh) was mounted in a retractable turret underneath the fuselage of the Coastal Command Wellington aircraft. In the final approach to the U-boat target detected by the ASV, the Leigh Light was switched on and this enabled the depth-charge attack to be carried out visually. The first Leigh Light Squadron of Coastal Command (No 172) was formed early in March† and began operations over the Bay in June. Hitherto the Germans had been sceptical that the Coastal Command equipped radar aircraft were effective. Now this sudden illumination and attack during what had previously been regarded as 'safe' hours upset the U-boat crews so much that although the total number of night sightings during June and July was only about 20, by August the U-boats were no longer surfacing at night for recharging purposes, but during the daytime. This enabled the day forces of anti-submarine aircraft to be deployed in daylight and by the end of 1942 the shipping losses had fallen to 200,000 tons per month. In his diary Admiral Dönitz wrote that 'there being no defence against aircraft in the Bay of Biscay, the RAF can do as it pleases'.[4]

The German scientists quickly provided the U-boat crews with a successful counter-measure in the form of a simple radio receiver known as

† A second Leigh Light Squadron (No 179) was formed at the beginning of September (1942).

'Metox'. The fitting of this receiver in the U-boats began in August (1942) and enabled the U-boat crews to pick up the radiations from an ASV-equipped aircraft long before it was at the range when it normally operated its Leigh Light. Thus the U-boats could submerge when their receivers indicated that an ASV transmission was homing on to them. The density of the aircraft patrols was insufficient to make the U-boats submerge so frequently that it became a nuisance to them, and the pre-June 1942 conditions were rapidly restored. After the 'seasonal' decrease, in the autumn of 1942 shipping losses took a sharp turn upwards in January and February 1943, and showed signs of reaching unprecedented proportions.

The Development of Centimetre ASV

The incidence of the U-boat listening to the aircraft ASV transmissions had one obvious technical solution, at least as a temporary measure—namely, to make a major change in the wavelength of the ASV transmission. The first airborne trials in March 1941 with centimetre AI equipment have been described in Chapter 6. The system fitted in Blenheim N3522 used the spiral scanner and although essentially designed for air-interception operations it was natural that an early opportunity should be taken of testing the ability of the equipment to detect surface ships and submarines. The first submarine tests were made in April against *HMS Sea Lion* and again in August against *HMS Sokol*. These tests, made mainly with Edwards and Downing as observers established that even if only the conning tower was exposed, the submarine could be located accurately and reliably.[5]

Soon after the April 1941 trials TRE made a formal application (June 1941) to DCD (Director of Communications Development) for authority to develop a centimetre airborne radar specifically for use against ships and submarines. Relative to the centimetre AI system this ASV development made rather slow progress. The references in Dee's diary have a despairing note. Although the fundamental features of this experimental centimetre ASV were identical with the AI, the first flight in a Wellington aircraft did not occur until December 1941 and in January 1942 Dee makes the exclamatory diary entry that 'ASV saw *Titlark* at 12 miles'.[6]

A separate group had been established in TRE to deal with the evolution of the centimetre ASV system and contracts were placed with Messrs. Ferranti and Metropolitan Vickers for the manufactured version. By chance, this development coincided with that of H_2S and during the summer of 1942 two essentially similar systems were being processed in TRE—ASVS with Ferranti/Metro Vic and H_2S with EMI/Metro Vic and Nash & Thompson. The H_2S system had a very high priority arising from the political and operational demands described in previous chapters. The ASVS system had no such priority. The ASVS group in TRE had not been formed in Dee's division and he repeatedly drew attention to the clash. Although this enhanced the natural competitiveness between the TRE groups the clash was never effectively acknowledged inside TRE until a measure of sanity was enforced from outside in the late summer of 1942.

As described in Chapter 15 soon after the crash of the Halifax, Renwick became responsible in the joint posts of Controller of Communications in the Air Ministry and Controller of Communications Equipment in MAP for H_2S and reported directly to the Secretary of State for Air and the Prime Minister. My initial contacts with him and Sayers occurred at Defford on 28 June 1942 and it was inevitable that they quickly became aware that two essentially similar centimetre systems were being developed by separate TRE groups with different firms. My first reference to their impact on this split is in the diary note for 26 September 1942.

> Yesterday [25 September 1942] Dee and I were in London to a Renwick meeting + 3 others. TRE is going to be clamped on by this reorganisation at the top. Holt-Smith and Fry [responsible for ASVS] were made to toe the line and the amazing agreement reached to use H_2S units for ASV and not boggle up Ferranti's with a separate system.

A few days later [30 September] Ferranti were instructed that their development of ASVS must cease and that they were to become daughter contractors to The Gramophone Company† for the manufacture of the system which now became H_2S/ASV. In that way at some of the most tense moments in the H_2S development the Holt-Smith/Fry ASVS group was disbanded and the part of the group dealing with ASVS was handed over to me to assist with the development and fitting of the ASV version of H_2S in Wellington aircraft of Coastal Command. With this change F C Thompson‡, who had been a member of the ASVS group, became an important senior member of my own group and in particular became our design authority for the production of the H_2S/ASV units at The Gramophone Company.

Relations with Coastal Command

The uneasy relations between Coastal and Bomber Commands were greatly exacerbated by the decision to use the H_2S equipment for ASV. The crash

---

† For the relationship of The Gramophone Company to EMI see footnote p 134 Chapter 14.
‡ F C Thompson had joined TRE in the summer of 1940. In the autumn of 1940 he was seconded to AA Command to assist with the commissioning and calibration of the GL equipment. In the spring of 1942 he was recalled to TRE for work with the ASVS group.

programme to produce a few Mark I H$_2$S systems by the end of 1942 was merely beginning at the time of this decision and Bomber Command fought bitterly to prevent this diversion of the scarce equipment they needed. Furthermore, Harris and his senior staff argued that far more damage would be inflicted on the U-boats by bombing the U-boat pens in the French West Coast ports than by attempting to sink them at sea.

For their part the senior officers of Coastal Command were very greatly annoyed by the decision to cancel the development of the system being undertaken specifically for their use. The C-in-C Coastal Command, then Sir Philip Joubert de la Ferté, visited TRE and announced that, in any case, he did not believe in ASV. More practically, the Chief Signals Officer, Group Captain Chamberlain, demanded a demonstration flight with this 'new system'. Greatly harassed we made a hasty arrangement for one of my group, I Beeching, to demonstrate the performance of an H$_2$S system fitted in a Stirling, over the sea and coastline. The results were excessively poor (discovered later to be due to the fact that we had not changed the tilt of the scanner). Those present at luncheon in the Defford mess after this flight are unlikely to forget the great irritation of Chamberlain.

The Coastal Command hierarchy would not face the reality that they *could* have the H$_2$S/ASV system almost immediately whereas they had no hope of getting their own ASVS system for many months. Further, the logical (to us) procedure of fitting a few extra four-engined bombers to be used by Coastal was violently opposed by both Coastal and Bomber Commands for different reasons.

Dec. 8, 1942. Great upheavals at work owing to enormous high level discussions about whether to use our heavies [i.e. the four-engined bombers]. We fought a great battle but got no backing from Joubert and were told to get on and fit the Wimpeys [Wellington aircraft].

It would have been logical to fit the scanners in the underbelly position of the Wellingtons as we had done for the four-engined bombers but that position was occupied by the retractable Leigh Light and the only feasible alternative was to use the nose position. However, this meant removing the forward guns, and since the Leigh Light had already replaced the under-turret guns there was quite naturally further argument and dispute. Eventually a perspex 'chin' was fitted under the nose of Wellingtons (see figure 18.1), but the scanner had to be re-designed to an aperture of 28 inches instead of 36 inches. The chin position prevented complete all-round-looking and caused a blackout in a 40 degree sector directly behind the aircraft.

By December 1942 we had fitted two Wellington VIIIs (LB129 and LB135) with the modified H$_2$S equipment and with the scanner in the nose position. There seemed to be endless teething troubles and the lack of enthusiasm and general antagonism of Coastal Command increased our difficulties. They demanded that we should include various beacon facilities and this caused delays and was of insignificant value compared with the major issues at stake. Many of the modifications were carried out on the top floor of the Malvern College Preston Laboratory by a few members of my group and, by the end of January 1943, we had sent two Wellington XII aircraft fitted with the H_2S/ASV and one of our prototypes to the Coastal Command Squadron based at Chivenor in North Devon. By mid-February we had sent seven Wellingtons to Chivenor and there were a dozen there by the end of the month. O'Kane, who had so successfully given the initial training in the use of H_2S to the Pathfinder Force navigators of 8 Group, was re-directed to Chivenor where he arrived on 8 March to help with the training of the Coastal Command navigators.

Figure 18.1 The Coastal Command Wellington LB129—the first to be equipped with 10 centimetre ASV, showing the housing for the scanner in the chin position.

Whereas the Bomber Command aircraft were operating at 20 000 feet the normal Bay patrols of Coastal Command were at 2 000 feet. The Wellingtons arrived at Chivenor with the normal dipole feed used in the H_2S scanners. In order to improve the coverage and give better performance when flying at the 2 000 foot level we decided to replace the dipole by a waveguide feed. The RAF Corporal Calcutt secured a machine that enabled him to mould the necessary special connectors and the waveguides were fitted when the Wellingtons were already at Chivenor.

The Coastal Command Officers exhibited negligible interest and with O'Kane and seven members of my group at Chivenor maintaining and giving instructions in the use of the equipment, one of my senior staff, Richard Fortescue, most generously offered to don RAF uniform and fly over the Bay as operator of the ASV.

March–April 1943

Although the aerodrome at Chivenor was in the midst of the attractive region of North Devon and close to popular peacetime seaside resorts my memory of the March days of 1943 are not pleasant ones. On the night of 1 March two of our 10 centimetre ASV-equipped Wellingtons took off from Chivenor for their first Bay patrol. Fervently believing that major issues were involved we had connived with the diversion of H₂S units from Bomber Command to their great annoyance and in the face of opposition we had staked nearly half of our group and much of our reputation to get these two aircraft operating over the Bay.

I have no record of the name of the Station Commander at that time but he had absorbed the antagonism of his superiors and vented a good measure of it on me as we waited for the return of the patrol. His office was in a typical pre-fabricated single storey building of that wartime period. The concrete floor was covered with brown linoleum and the yellowish paint was flaking off the walls. It was not a good place to wait for the grey cold dawn of 2 March. His faith in our equipment was entirely lacking. For him it was another nuisance. 'What good do you think that gadget of yours is going to do out there. You ought to fly out and look for yourself at that great feature-less expanse then you would realise that your rotating thing would never stand a chance of catching a sub.'

I would gladly have flown out and looked for myself but it had been nearly impossible to arrange for Fortescue to go and if I had been there my eyes would have been glued to the cathode-ray tube—as Fortescue's were. It was as well that he was looking at the tube. At last he returned. No, there had been no submarines but they had been attacked by a German fighter and he had been able to give the pilot instructions for evasive action. The crew of that Wellington spread the news and the tensions of our presence on the aerodrome began to ease.

For two weeks the patrols continued without incident. With seven specialists from my group looking after the equipment the serviceability was good and then, during the night of 17 March, the equipment in Wellington H538 saw an echo from a U-boat at a range of nine miles. The attack on the U-boat was foiled by the jamming of the Leigh Light. The next night H538 obtained another contact at a range of seven miles and this time the U-boat was attacked with six depth charges. The pilot's log read: 'Both times the submarine was fully surfaced and under way, showing no signs of suspecting attack'. 'March 27th. Chivenor had a good haul in the Bay this week—4 good attacks!'

The centimetre ASV—now coded as ASV Mark III in this 172 Squadron based at Chivenor, made 13 sightings before the end of March and another 24 in April. The U-boat crews did not relish these night attacks and Dönitz ordered his crews to stay on the surface and fight it out with the aircraft, not only on the Bay transit routes but around our convoys.[7] Now, long-range aircraft operating from both sides of the Atlantic closed the mid-Atlantic gap, and a grim slaughter of the U-boats ensued. During the decisive months of April and May 1943, 56 U-boats were destroyed and at the end of May the Naval Staff 'noted with a relief that can still be felt today the sudden cessation of U-boat activity which occurred on or about the 23rd of May'.[8]

The few 10 centimetre ASV-equipped Wellingtons in 172 Squadron had, within a few weeks, transformed the strategic situation to such an extent that in May 1943 every U-boat crossing the Bay, on average, suffered one attack. The shipping losses in the North Atlantic had reached 400 000 tons in March but, instead of rising to the unprecedented levels feared by the War Cabinet they then fell abruptly and by August were less than 100 000 tons per month (figure 18.2). In a radio broadcast Hitler complained that 'the temporary setback to our U-boat campaign is due to one single technical invention of our enemies'. The dramatic nature of this change is evidenced by Roskill's[7] comment on these events in his official history of the war at sea. 'For what it is worth this writer's view is that in the early Spring of 1943 we had a very narrow escape from defeat in the Atlantic'.

Concerning the introduction of the 10 centimetre ASV over the Bay in the Spring of 1943 Roskill states[9] that

> the Air Ministry ordered the diversion of the first forty sets [of H_2S] to the Leigh Light Wellingtons. This however, could only be a stop gap and was unlikely to be wholly satisfactory because the set had been developed for a different aircraft employed on a different function. The only adequate solution was to get the new sets from the USA, where they were now being fitted to Liberators.

This comment of Roskill's is somewhat misleading. The 'first forty sets' of the 200 crash programme production of H_2S had been fitted in the Pathfinder Force of Bomber Command as described in Chapters 16 and 17 and those used by 172 Squadron of Coastal Command over the Bay of Biscay in March–April 1943 were the few modified in TRE as described earlier in this chapter.

As regards the American centimetre equipment in the Liberators, D M Robinson had gone to the USA (as described in Chapter 16) in July 1941 to supervise the fitting of lend–lease aircraft with centimetre ASV. In March 1942 he returned with the first Liberator equipped with a prototype 10 centimetre ASV made in the Radiation Laboratory. Trials were carried out against a British submarine in Loch Neagh in Ireland and as a result of these

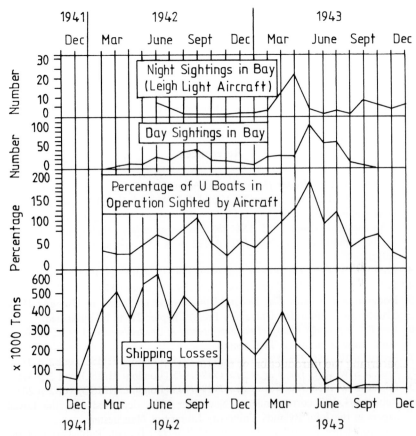

Figure 18.2 Merchant shipping losses and U-boat sightings by aircraft in the Bay of Biscay, 1942–43, illustrating the dramatic change after the introduction of 10 centimetre ASV in the spring of 1943. The curve marked 'percentage' is based on the number of sightings of the estimated number of U-boats operating in the Bay of Biscay and shows that the two upper curves are tactically real and not merely due to an increase in the number of U-boats operating.

trials Robinson returned to the USA to supervise the installation of more Liberators with the American 10 centimetre ASV.[10] These prototype ASV sets were designated DMS-1000 and were first used operationally over the Atlantic and Bay of Biscay early in 1943. These early Liberators did not have Leigh Lights and their major impact occurred later in the summer in closing the Atlantic gap when they were fitted with the production version of the American 10 centimetre ASV made by the Philco Corporation. The details of the development and initial operations have been given by Guerlac[11] who writes that the 'barrier patrol' of Coastal Command across the Bay of Biscay was

> maintained at night largely by means of a few Wellington airplanes, fitted with Leigh Lights and 10 cm radar, of whose existence the Germans had as yet no knowledge. This early 10 cm ASV was all preproduction equipment; modifications of the British blind bombing set (H_2S)...[12]

References

1   Bowen E G 1987 *Radar Days* (Bristol: Adam Hilger)
2   Smith R A., Hanbury-Brown R, Mould A J, Ward A G and Walker B A TRE monograph 1945.
    An abridged version 1985 *ASV: the detection of surface vessels by airborne radar* has been published in *Proc. IEE* A **132** pt A 359 (special issue on historical radar)
3   Roskill S W 1956 *The War at Sea 1939–1945* (London: HMSO) vol 2 p 111
4   Roskill S W 1956 *The War at Sea 1939–1945* (London: HMSO) vol 2 p 112
5   Hodgkin A L *Chance and Design: Reminiscences of Science in War and Peace* (Cambridge: Cambridge University Press) Chapter 21, to be published
6   Dee P I *A personal history of the development of cm. techniques AI, ASV, H_2S and AGLT* (copy is in Lovell TRE record)
7   Roskill S W 1956 *The War at Sea 1939–1945* (London: HMSO) vol 2 p 371
8   Roskill S W 1956 *The War at Sea 1939–1945* (London: HMSO) vol 2 p 376
9   Roskill S W 1956 *The War at Sea 1939–1945* (London: HMSO) vol 2 p 205
10 Robinson/Lovell correspondence November 1990, deposited with the Lovell papers in the John Rylands University Library of Manchester.
11 Guerlac Henry E 1987 *Radar in World War II* (Tomash Publishers and the American Institute of Physics) Chapter 12 pp 321–5
12 Guerlac Henry E 1987 *Radar in World War II* (Tomash Publishers and the American Institute of Physics) Chapter 36 p 720

Chapter 19

The Impact of the German Naxos on Centimetre ASV

The emergency conversion of the early 10 centimetre H_2S to operate in Coastal Command as ASV was stimulated by the need to evade the U-boats' use of the listening device Metox which detected the 1½ metre ASV transmissions from an approaching aircraft. As described in the previous chapter, in the spring months of 1943 this had a dramatic tactical effect on the Allied air war against the U-boats.

We believed that this would be merely a temporary relief because there seemed no reason why the Germans should not immediately detect the change in wavelength of the ASV transmissions and modify their listening receivers accordingly—particularly since one of the 10 centimetre H_2S Bomber Command aircraft had been shot down near Rotterdam in February (see Chapter 29). With this prospect in view we immediately began the preparation of a 3 centimetre ASV equipment which would give at least another temporary relief. During the late spring of 1943 we developed an appropriate 3 centimetre ASV for fitting in Coastal Command Wellingtons in parallel with the 3 centimetre H_2S development (see Chapter 22)—coded ASV Mark VII. In addition we prepared a modified scheme for the 10 centimetre ASV. This used a more powerful S-band magnetron giving 200 kilowatts in the pulse and an attenuator—ASV Mark VI. The idea was that as soon as the radar echo from the U-boat was seen the attenuator would be operated to decrease the radiated power. Thus with the aircraft still homing on to the U-boat, a listening receiver in the U-boat would detect a steadily decreasing signal giving the impression that the aircraft was, in fact, receding. As a further improvement we had proposed to add a lock–follow system—ASV Mark VIA.

Surprisingly the Germans were slow in their reaction and their tardiness in realising that the ASV planes were transmitting on 10 centimetres instead of 1½ metres has never received a full explanation. As will be described in Chapter 29, the German intelligence officers and scientists discovered without delay, from the bomber that crashed near Rotterdam during the night of February 2–3 (1943), that it carried a system capable of generating high powers in the centimetre waveband, but a considerable time elapsed before the Telefunken Company succeeded in re-creating a working system from H_2S units recovered from crashed bombers. The immediate emphasis was then on the development of a simple receiving system, Naxos, for the detection of these centimetre transmissions and the major interest was on the use of Naxos in the German night fighters to assist them in locating and attacking the RAF bomber stream. It was mid-May (1943) when Admiral Dönitz reported to Hitler:

> We are at present facing the greatest crisis in submarine warfare, since the enemy, by means of location devices makes fighting impossible and is causing us heavy losses.[1]

The German U-boat losses had risen to 30 per cent of those at sea and the surviving U-boats were withdrawn from the north Atlantic. Dönitz attributed this defeat to the inability of the U-boat search receivers to detect the radar transmissions from the Coastal Command aircraft. As Roskill points out[2] Dönitz was correct 'up to a point' in this opinion. Of course, many other factors contributed to the defeat of the U-boat campaign during those months[2]; nevertheless when it was realised that the technical difficulties involved in producing a centimetre radar for the RAF anti-submarine aircraft had been overcome, a form of Naxos was then commissioned for use in the submarines to give warning of the approach of the Coastal Command centimetre ASV aircraft—that is there was a delay of nearly three months after the first use of centimetre ASV over the Bay of Biscay before action was taken by the Germans to replace their 1½ metre Metox listening receiver by the 10 centimetre Naxos. At first it appears that the U-boat Commanders thought that the Coastal Command planes must be detecting re-radiation from the U-boat Metox receivers and homing in on this and not by means of ASV. Indeed, Roskill[3] states that when Dönitz summed up his experiences at the end of August 1943 he believed that this was the case but 'the conclusions at which he had arrived were quite wrong, for he had been misled by the scepticism of his own scientists regarding the possibility that we might be using centimetric radar ... and by the false information given by the pilot of a Coastal Command aircraft which had crashed'. That officer told his interrogators that our aircraft were able to 'home' themselves on to U-boats by means of emission from the German search receivers; and, after ascertaining that this was in fact technically possible, the Germans accepted the pilot's assurances as gospel. In any event the delay in their understanding that the Coastal Command ASV was now radiating on a shorter wavelength illustrates the lack of close liaison between the scientists and the operational staffs. It is inconceivable that such a delay could have occurred with the Allies where the scientific/operational liaison was so close.

The consequence of this delay was that by the autumn of 1943 when evidence arose that the Germans were now listening to the 10 centimetre ASV transmission† we had completed our development of ASV Mark VI and VII. Coastal Command wanted this equipment fitted in their Wellington Mark XIV aircraft and just as we were acquainted with evidence of the Germans listening to the 10 centimetre ASV a most extraordinary circumstance arose.

> Sat. Oct. 30th (1943) ... the chief event is the trouble over a Wellington XIV for our ASV VI programme. After weeks of frustration in getting a T.I. [Trial Installation] plane and all laid on, a meeting was at last organised last Friday week at Defford and to my great sorrow we again spent the time arguing whether there was an operational requirement or not! Only the night before Rowe had rung me with a message from the C in C saying that the Germans were undoubtedly listening to Mk 3.

That frustrating meeting at Defford on 22 October had immediate consequences. Air Marshal Sir John Slessor had succeeded Joubert as C-in-C Coastal Command and on Saturday afternoon 23 October he came to TRE to make sure that everything possible was being done to meet this new threat. When he discovered that we had completed our development but that some unknown persons in London had successfully sabotaged the trial installation for many months past he became very angry indeed. The next day—a Sunday—he delivered a five page foolscap letter to the Under Secretary of State for Air—for the attention of the Deputy Chief of Air Staff —which caused 'enormous consternation'. This letter[4] set out in great detail the report which he had made to the War Cabinet Committee on anti-U-boat warfare‡ on 22 March (1943) drawing attention to the urgent need for further changes to the 10 centimetre ASV since he believed 'it was only a matter of time before the Germans would be able to listen to 10 cm ASV'. After having detailed his subsequent statements he referred to the completion of the development at TRE and asked for disciplinary action to be taken against those responsible for the delay in providing the trial installation Wellington. He continued that this was 'either crass stupidity or pettifogging obstructionism of the worst kind I have ever encountered in this war' and ended by stating that 'no one but a congenital idiot would imagine that it was not necessary to lay on Column 7 [i.e. the aircraft and fittings] when Column 9 [i.e. the equipment] was already provided'.

This was an extraordinary letter couched in undiplomatic language from

---

† Roskill's statement[3] that 'Not until the beginning of 1944 did it dawn on the Germans that our sets were working on the ten centimetre waveband' is at variance with the operational evidence presented to TRE by the C-in-C Coastal Command in the autumn of 1943.

‡ The Anti U-boat Committee held fortnightly meetings at 10 Downing Street under the Chairmanship of the Prime Minister.

a C-in-C who was exasperated and led to instant action in the Air Ministry. Three days later, on Wednesday afternoon 27 October, I was summoned to an urgent meeting in the War Cabinet office. At 8 am on that morning as I was about to leave Malvern, Dee telephoned to say:

> for heavens sake watch my step as every person in MAP and AM was going around white lipped wondering if he [Slessor] was going to be court-martialled.[5]

If Slessor's letter was extraordinary the meeting was even more so. It was the first and only time at which I had witnessed a direct confrontation between the hierarchy of Bomber and Coastal Command. The Chairman was the Deputy Chief of Air Staff, Air Marshal Sir Norman Bottomley. On his right was the C-in-C Coastal Command (Slessor) with his staff. On Bottomley's left sat CCE (Renwick), VCCE (Watson-Watt), Saundby (the Deputy C-in-C Bomber Command), Bennett, and a mass of Air Ministry Operational Staff. I was placed at the bottom of the vast table with the Air Ministry Director of Radar and the Director General of Signals.

As described in Chapter 22 Bomber Command had been promised 200 H_2S 3 centimetre sets by the end of the year. The issue which DCAS placed before the meeting as a result of Slessor's letter was whether 50 of these 3 centimetre sets should be diverted to Coastal Command for ASV use. Saundby and the Bomber Command group were enraged and argued that the most efficient use of the 3 centimetre H_2S would be for them to use it to destroy the U-boats in their pens on the West Coast of France. After a tremendous session of argument and counter-argument across the table Slessor eventually managed to persuade DCAS to allocate enough of the 3 centimetre equipment for one of his squadrons and Renwick, who was greatly annoyed with the whole affair, promised to make good the deficiency to Bomber Command as soon as possible. As the papers were being folded Saundby said to DCAS:

'I will report the decision to my C in C but I know that he will be much displeased'.

I returned to Malvern to be met at the gates of TRE by Rowe.

'You haven't done very well Lovell'.

'Why not? Coastal are being given enough for one squadron'.

'No, when Saundby reported this decision to Harris he immediately telephoned the Prime Minister and got the decision reversed'.

So Coastal Command did not secure any diversion of the 200 crash pro-gramme 3 centimetre ASV from the Bomber Command programme. The 3 centimetre ASV Mark VII did eventually materialise but the priorities and effort directed to this steadily decreased as the continued failure of the U-boat campaign became evident. In the meantime, throughout this saga, Wing Commander Leigh (of the Leigh Light) constantly agitated for the lock–follow version of the high powered 10 centimetre system (ASV Mark VIA) so that his Leigh Light could be locked on to the target in synchronism with the ASV scanner. It was mid 1944 before we managed to produce an experimental version of ASV Mark VIA, (figure 19.1) but by that time the Allies had captured the West Coast ports of France and the Battle of the Bay of Biscay was already history.

Figure 19.1 The ASV Mark VIA display unit in the navigator's cabin of a Wellington XIV.

At the time of the confrontation between Coastal and Bomber Command about our 3 centimetre H_2S in the autumn of 1943, I did not know that this was another aspect of a bitter dispute that had arisen six months earlier. On 30 March 1943 the Admiralty had launched a paper prepared by Blackett (then Director of Naval Operational Research in the Admiralty) directly to the Anti-U-Boat Committee. In this paper, prepared without any consultation with Slessor, Blackett estimated the number and type of aircraft required to have a decisive effect against the U-boats in the Bay of Biscay.

He proposed a force of 260 heavy aircraft which would have required a diversion from Bomber Command to Coastal Command of 190 heavy bombers. Slessor was greatly annoyed: 'The operational research scientist has no stronger supporter than I ... but they must stick to their lasts. Statistics are invaluable in war if they are properly used ... But the Bay offensive was a battle ... and nothing could be more dangerously misleading than to imagine that you can forecast the result of a battle or decide the weapons necessary to use in it, by doing sums ... and in fact it (i.e. the Battle of the Atlantic) was won with a fraction of the number of long range aircraft postulated in the scientific study ...' Slessor complained to the First Sea Lord about this 'slide-rule strategy of the worst kind'. He was convinced that he did not want 190 heavies diverted from Bomber Command but 'aircraft of the right type, with the right sort of radar equipment and with crews trained in the right way— and I wanted them quickly. *Now* was the time when we wanted to kill the U-boats, while we had the bulge over them with the 10 cm ASV, and I was relatively uninterested in what would be happening in 6 months time'. To the Chief of Staff, Slessor wrote: 'Summarising my objections to the principle of strategy by slide-rule, I urged that the problem should be tackled from a less scientific but more practical angle. As long as the enemy continues to present us with fat profits in the convoy area, we should make the best use of our opportunities there. At the same time we should build up our strength in the Bay as quickly as we could'.[6]

The question of whether Blackett or Slessor was right in this conflict remains difficult to assess. Blackett's paper had been based on calculations made by E J Williams, a member of his former operational research group in Coastal Command who had followed him to the Admiralty. But the calculations were based on the conditions existing in the Bay of Biscay before the impact of the 10 centimetre ASV had transformed the tactical situation in March/April 1943.

Probably both Blackett and Slessor were substantially correct although the divergence seemed wide at the time. Although Slessor states that the battle was won with 'a fraction' of the aircraft demanded by Blackett's paper, the numbers diverted eventually were 70 against the 190 suggested in the paper; and, of course, the Liberators from the U.S. were soon to play a decisive role.[7]

References

1   Roskill S W 1956 *The War at Sea 1939–1945* (London: HMSO) vol 3 pt 1 p 15
2   Roskill S W 1956 *The War at Sea 1939–1945* (London: HMSO) vol 3 pt 1 p 16
3   Roskill S W 1956 *The War at Sea 1939–1945* (London: HMSO) vol 3 pt 1 p 32–3
4   A copy of this letter Reference S.7012/11/C-in-C dated 24 October 1943 headed *Immediate and Most Secret* is inserted in Lovell, TRE record p 14
5   Lovell, extract from diary note, October 30, 43
6   This extract is from the Royal Society biographical memoir of P M S Blackett— Lovell, *Biog. Mem. R. Soc.* 21 1 (1975), in which further details of this conflict can be found. This is based on Slessor's own account in *The Central Blue* (London: Cassell), 1956.
7   Lovell 1975 *P M S Blackett a biographical memoir* in *Biog. Mem. R. Soc.* 21 p 64 and Slessor J 1956 *The Central Blue* (London: Cassell)

Chapter 20

H_2S on Tank Landing Craft

The H_2S equipment so far discussed had been manufactured at EMI and RPU on a crash programme basis, limited to 200. The main production was scheduled to commence in May 1943 and it was evident that because of the high priority attached to H_2S this would be the main source of 10 centimetre equipment for some time. The prospect of the availability of large numbers of the H_2S system had other consequences which first affected me during the tension of the March 1943 events at Chivenor.

On 21 March an officer from Combined Operations, Lt H F Short and a civilian, E M Gollin, from the operational research section at Combined Operations, came to TRE to enquire into the possibility of using H_2S or a modification thereof for navigating and ranging on tank landing craft. They departed and there was silence for over two weeks and then:

> April 14, 1943. Just back from Portsmouth—beginning of new flap for fitting our gear on tank landing craft. Dee came and Chisholm—a fellow from Taylor's crowd.

The reference to 'Taylor's crowd' was to a group led by Dennis Taylor working in another part of TRE quite separate from Dee's division. They were supposed to have the TRE responsibility for anything to do with radar on ships which was not being handled by ASE (Admiralty Signals Establishment). However, neither Taylor nor ASE had the centimetre system that we now had for H_2S and the job of equipping a tank landing craft with a modified H_2S quickly landed on my desk.

The landing craft in question known as a LCT was, in fact, a rocket-firing ship and our journey to Portsmouth on 14 April was to make a survey to discover how we would have to modify and fit the H_2S units. As far as we were concerned it was an easy task and on 19 April one of the RAF mechanics, Hinckley, attached to our group drove the modified set of H_2S to Portsmouth. This was one of the early main production versions—H_2S Mark II—and the only modifications needed concerned the range scale on the PPI which we changed to give two options—3 nautical miles and 20 nautical miles. The dockyard had promised to build a tower on the deck of the LCT so that the scanner could be mounted 36 feet above sea-level. They were slow in doing so and although we provided the perspex housing for the scanner it was 9 May before the installation was ready for trial.

9 May 1943 was a Sunday and a very beautiful one with the waters of the Solent glistening in the sunshine. As well as Short and Gollin who originally came to see us, Commander R T Paul from Combined Operations was on board.† There was not yet an official code-name and we had christened the device 'Cobweb'.

> May 9, 1943. Cobweb has been a howling success ... Combined Ops teared their hair out with delight

Combined Operations had determined the purpose of these trials—(a) to examine the range and efficiency of coastline painting with particular reference to the suitability of the device as an aid to running on to a selected part of a coastline; and (b) to examine the accuracy of the range measurements on selected objects and coastlines. (An accuracy of 50 yards at 4000 yards was said to be desirable).

At the end of the trials on 9 May there was no question about the success of the H_2S in meeting these requirements. Low-lying coasts were visible on the PPI at ranges of 10 to 15 miles and the south entrance to Portsmouth harbour was clearly defined. The harbour indentation at Bembridge on the Isle of Wight was clearly painted on the tube before it could be seen visually. The H_2S ranges on the Nab Tower, various forts and lightships and on coastlines corresponded with those determined by optical measurements from the bridge of the LCT. The sea was fairly rough and although the LCT rolled severely this did not affect the PPI presentation.

In the evening we left our H_2S equipment on the LCT and returned to TRE. A few days later the Chief of Combined Operations sailed with the system and we were instantly asked to modify 25 H_2S units. Sub Lieutenant Armstrong was sent to help and with four naval mechanics already at TRE on a radar course the job was soon completed. By 2 June six LCTs were equipped with the modified H_2S and ready for sailing.

We did not know that at that time Churchill was in North Africa discussing with the Americans the plans for the invasion of Sicily and Italy.‡ Neither did we know that the six LCTs equipped with H_2S now officially coded Scent Spray, were destined to take part in that invasion in the early hours of 10 July. Early in June the task of fitting LCTs with 10 centimetre radar passed from our hands to ASE. A year later in the invasion of Europe it had become the Admiralty Type 970 and only the photographs of the PPI taken from an LCT off Cherbourg on D-Day reminded me of the pleasant Sunday on the Solent in May 1943.

† Also on that trip there was Squadron Leader Gilfillan from the Air Ministry, Hinckley, Chisholm and myself from TRE.
‡ He left America in a Boeing flying boat accompanied by President Roosevelt on 26 May and spent eight days in discussion with the Americans in Algiers and Tunis.

Chapter 21

The Summer of 1943—
Destruction of Hamburg

A few days after the first use of H_2S at the end of January 1943, Harris received a new directive. At the Casablanca Conference it was agreed on 21 January 1943 that the conduct of bombing operations should be the responsibility of the Combined British and American Chiefs of Staff. On 4 February Harris and General Eaker (see Chapter 23) received the directive that their primary objective should be:

> the progressive destruction and dislocation of the German military, industrial and economic systems, and the undermining of the morale of the German people to a point where their capacity for resistance is fatally weakened. This is construed as meaning so weakened as to permit initiation of final combined operations on the continent.†

Shortly after Harris received this directive the euphoria arising from the initial successful raids using H_2S began to evaporate. There were increasing complaints about the gaps and fades in the PPI picture as the bombers approached the target. This old problem, troublesome enough at 15 000 feet, became greatly aggravated with the increase in operational height of the bombers to 20 000 feet. Ramsay and Hillman carried the major burden of flight-testing our attempts to rectify this difficulty. Significant improvements were made in the vertical polar diagram by using a different dipole feed arrangement to the scanner and by introducing a baffle plate (figures 21.1, 21.2). Then in April we abandoned the dipole feed in favour of a shaped waveguide and this led to a major improvement in the smoothness of the presentation during the run up to bomb release (figure 21.3). With the urgent backing of Bomber Command formal application was made on 19 May for the manufacture of conversion kits. By July the H_2S Pathfinder aircraft were converted and these led the main bomber force in the devastating attacks on Hamburg towards the end of July.

† The full directive with details of the primary targets to be attacked is reprinted in reference 1.

Figure 21.1 March 1943 10 centimetre H_2S with modified dipole feed arrangements plus baffle. The towns are Cheltenham and Gloucester with the aerodrome between them. Although a great improvement on the original system (compare figure 16.1) the ground clutter at short range was still troublesome. (The straight line is the heading marker.)

Figure 21.2 The scanner, and TR box in a Halifax. The scanner shown here is using the original adjustable 'baffle plate' and dipole feed—H_2S Mark I.

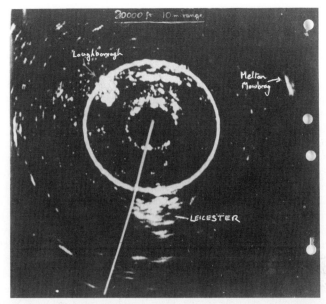

Figure 21.3   April 1943. The major improvement in 10 centimetre H_2S by use of a waveguide feed to the scanner. This PPI photograph was taken in a Lancaster flying at 20 000 feet. The straight line is the heading marker. The circle is the 10 mile range. Leicester and Melton Mowbray are clearly defined.

For the previous three months as the nights shortened the main targets for Bomber Command had been in the Ruhr. The targets were within range of Oboe and had been marked with high accuracy. The success of these Oboe marked raids strengthened the disparate elements, both within TRE and in the operational command, that were antagonistic to H_2S. The mass production of H_2S for the main bomber force had commenced in May. Apart from minor display improvements, facilities for using beacons and some re-arrangement of the controls this H_2S Mark II was no improvement on the crash programme H_2S Mark I so far fitted in the Pathfinder Force. Indeed there were occasions when, incredibly, it seemed possible that the whole H_2S fitting programme would be cancelled.

For us in TRE, for Saward and many of his pro-H_2S colleagues in Bomber Command, the massive raids planned on Hamburg began to emerge as a crucial test for the future of H_2S. In the raids earlier in the year relatively small numbers of the Pathfinder Force aircraft had been equipped with H_2S. Now almost the whole of Bennett's force had H_2S and several had been modified to take the new waveguide feed to the scanner, and considerable numbers of the main force also had H_2S. Saward[2] has described Project *Gomorrah*, in which, over four nights from 24–25 July to 2–3 August 1943, between 700 and 800 aircraft per night dropped over 9 000 tons of high explosive and incendiary bombs on the city.

> In July 1943, Hamburg became an urgent target to attack, with much pressure for its elimination from the Admiralty. Hamburg was the second largest city in Germany, with a population in excess of a million and a half. Its shipyards were the most extensive in Europe, housing many ships and U-boats under construction. The greater part of its shipbuilding yards, including the famous Blohm and Voss yards, had been given over to the building and assembly of U-boats, and they were responsible for some 45 per cent of the total output of German submarines. Hamburg was also the largest and most important port in Germany. It contained 3,000 industrial establishments and 5,000 commercial companies, most of which were engaged in the transport and shipping industries. In addition there were major oil and petroleum refineries, the second largest manufacturer of ships' screws, the largest wool-combing plant in Germany, and various manufacturers of precision instruments, electrical instruments, machinery and aircraft components ... Whilst Hamburg was outside Oboe range and therefore the Pathfinders could not take advantage of this very precise marking device, Harris now had at his disposal excellent RDF aids for the plan of attack. H2S was, at this stage, fitted to a considerable portion of the main bomber force as well as to the Pathfinders, and since Hamburg was an ideal H2S target, being a seaboard town and therefore easily identifiable from the map-like display of coastlines and rivers provided by H2S on its cathode ray screen, not only could there be certainty of exceptionally accurate marking, but the main force could be sure of following up the Pathfinders with equal accuracy ... The route was out over the North Sea to a position exactly fifteen miles north-east of Heligoland, which was the point at which the bomber force was to turn in towards its target. With Gee working perfectly up to these distances the accuracy of navigation was excellent, timing perfect and concentration maximal. Both Pathfinder and main force aircraft then had the aid of their H2S to identify the coast in the neighbourhood of Cuxhaven and the Rive Elbe as it unfolded itself on the screen as far as Hamburg, and finally as it revealed the bright fingers of light of the dock area ... It was more than a week before the true extent of the damage could be assessed, photographic reconnaissance being impossible for the first six days after the last attack due to the continuous presence of a vast pall of smoke over the entire city. When photographic evidence was available it showed that 75 per cent of the city had been razed to the ground. More than ten square miles of built-up area had been eradicated. Shipyards, industrial enterprises, commercial areas and residential areas had been wiped out, and city services were no longer able to operate.

In the official history of the strategic air offensive Webster and Frankland write of these raids:

> ... the last attack (2 August) came as a macabre climax to the great catastrophe which had overwhelmed the whole city, and for Bomber Command the victory was complete.[3]

Figure 21.4 H$_2$S PPI and control unit in the navigator's cabin of a Lancaster.

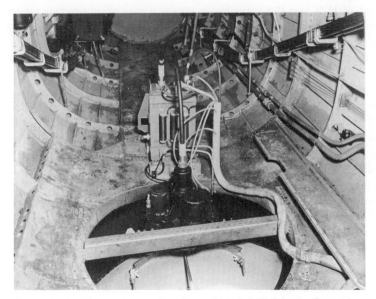

Figure 21.5 The scanner and TR box of Mark I H$_2$S in the Lancaster.

The full extent of the devastation of Hamburg in these raids and the effect on the German war effort became apparent after the end of the war. The report by the Police President of Hamburg dated 1 December 1943 concluded that:

> The Utopian picture of a city rapidly decaying, without gas, water, light and traffic connections with stony deserts which had once been flourishing residential districts had become reality.[4]

In the interrogation of Reich's minister Albert Speer (Minister for Armaments and War Production) on 18 July 1945 the following exchange occurred[5]:

> Question At what stage of the war did strategic bombing begin to cause the German High Command and Government concern and why?
> Speer The 1st heavy attack on Hamburg in Aug. 43 made an extraordinary impression. We were of the opinion that a rapid repetition of this type of attack upon another six German towns would inevitably cripple the will to sustain armaments manufacture and war production. It was I who first verbally reported to the Führer at that time that a continuation of these attacks might bring about a rapid end to the war.

However, as Saward stated, Hamburg being a seaboard town was an excellent target for H_2S. This did not apply to several other cities and, with the spread of H_2S into the main force, where it was often used by less highly trained operators than those in the Pathfinder Force there were some failures. The attack on Hanover for example on 2 September was a fiasco. H_2S was blamed for the failure although five nights later a second attack was successful. Indeed, it often seemed to me during those summer and early autumn months of 1943 that a bad raid was attributed to a 'useless H_2S' whereas no comment was made on the successful raids which could never have been made without the help of H_2S. On October 22–23 Kassel had been attacked by 550 aircraft yet it was three months before the Operational Research Section in Bomber Command circulated the plot of night photographs that revealed the precise marking of the H_2S Pathfinder Force aircraft and that 75 per cent of the main force H_2S bombing was within one mile of the target point. However, the urgent need for a major improvement in H_2S had been revealed by the disastrous attempts of Bomber Command to attack Berlin in late August and early September 1943.

References

1   Webster Sir Charles and Frankland Noble 1961 *The Strategic Air Offensive against Germany 1939–1945* (London: HMSO) vol 4 appendix 23 p 273 *et seq*

2 Saward Dudley 1984 *'Bomber' Harris* (London: Cassell) p 208 *et seq*
3 Webster Sir Charles and Frankland Noble 1961 *The Strategic Air Offensive against Germany 1939–1945* vol 2 (London: HMSO) pp 150–157
4 Webster Sir Charles and Frankland Noble 1961 *The Strategic Air Offensive against Germany 1939–1945* (London: HMSO) vol 4 appendix 30 p 310
5 Webster Sir Charles and Frankland Noble 1961 *The Strategic Air Offensive against Germany 1939–1945* (London: HMSO) vol 4 p 378

Chapter 22

H_2S on 3 Centimetres (X-band)—the Attacks on Berlin and Leipzig

By late August 1943 the nights were long enough for Bomber Command to stage a major attack on Berlin. They did so on 23–24 August, 31 August–1 September and on 3–4 September, with poor results and heavy losses of aircraft. In these attacks 1400 heavy bombers were used and 104 did not return. Apart from the losses H_2S was blamed for the poor results. On 5 September a depressed Saward came to see me bringing photographs of the H_2S PPI as Berlin was approached. The city was so large that the H_2S picture was very confused. On the 10 mile range scale (the 'bombing' scale) the responses covered the whole PPI and there were no clear outlines on which the navigator could make a pin-point of his position. Of those attacks on Berlin, Webster and Frankland wrote in their official history that, compared with Hamburg, Berlin 'possessed no salient H_2S characteristics'[1] and that:

> the revolutionary factor in the battle of Hamburg was the release of Window†, but the attack also signalised the most rewarding employment of H_2S. The Battle of Berlin on the other hand began and continued without corresponding new advantages.[2]

Dee was away from TRE—

Sunday, Sept.11th [1943] Saundby—deputy C in C bomber command was down yesterday. I have a hard job with Dee away standing up to the anti-H_2S barrage of the rest of the station. They keep trying to push crazy schemes like GH repeater beacons and 1½ m repeater Oboe as the thing to bomb Berlin in 6 weeks! We've started up a scheme of a few 3 cm Mosquitoes. It is a racket.

Three days later our scheme for 'a few 3 cm Mosquitoes' was transformed

† See Chapter 7 and footnote p 81.

180

into a desperate plan in consultation with Saward and Bennett. One wall of my office was completely covered with a large map of Europe. When Saward showed me the confused PPI pictures of Berlin I looked up at this map and saw the thin blue line of the Tegelersee and Wannsee stretching south west of the city. One could just imagine a suggestion of them on one or two of the better PPI pictures. That was with 10 centimetre H_2S. Surely on a 3 centimetre H_2S they would be clearly revealed on the PPI screen.

> The hope of having some 3–centimetre H_2S in November/December through official sources was, by September, a forlorn hope, and Bennett of the Pathfinders, now an Air Vice-Marshal, and Saward, had hatched a desperate plan with Drs Dee and Lovell to build six 3-centimetre sets at TRE and to install them in six Pathfinder Lancasters by the middle of November. Harris and Saundby, when told of the situation, had given their blessing. The only other person in the picture was Sir Robert Renwick, whose unofficial help Saward had sought. This support by Harris of what was clearly a private venture being conducted outside the official channels, was typical of his approach when he needed something urgently.[3]

At the time I wrote that this was the 'craziest plan possible'. Indeed it was so. I think only in the heat of battle can such plans be made and carried through in defiance of logic and official plans.

It was, of course, a natural step to develop the magnetron and associated techniques for use on a wavelength of 3 centimetres. This work involving Skinner's group in which the senior members were Atkinson, Ward and Starr had been working on the development of techniques for 3 centimetres in close collaboration with the 3 centimetre group of the Admiralty Valve Laboratory at Birmingham University under J Sayers. In the early autumn of 1942 the group fitted an airborne system in an American Boeing passenger aeroplane† that had been sent to this country in the summer of 1941 to demonstrate the American version of centimetric AI.[4] This equipment gave good results on coastlines, estuaries and similar features when flying at 2000 feet and production arrangements were in progress to meet a Fleet Air Arm requirement.

It was our view that the success of these demonstration flights at low altitude gave a completely misleading view of the problems that would be encountered in attempts to emulate this for H_2S where the altitude was 20 000 feet and a truncated scanner had to be fitted underneath the fuselage of the aeroplane. Unfortunately our pessimism about this was soon to be revealed in practical form. In January 1943 when the Pathfinder Force was in the final stages of preparing to use the 10 centimetre H_2S operationally,

† I am informed by J R Atkinson who was in charge of the 3 centimetre installation in this aeroplane that this was not the Boeing originally fitted with the American centimetre AI equipment as recorded in my record. He believes it was an executive passenger aeroplane already in this country.

J R Atkinson at work on the experimental 3 centimetre airborne system in an American Boeing passenger plane—autumn 1942. (Photograph by courtesy of J R Atkinson.)

we began to convert a Stirling bomber, then fitted with 10 centimetre H₂S, for 3 centimetre operation. We were then immersed in our attempts to fit 10 centimetre ASV into the Coastal Command Wellingtons as well as with the Bomber Command problems and only one member of the group, Beeching, could be allocated to the Stirling conversion. He managed to extract some 3 centimetre magnetron equipment from the Skinner/Atkinson group, and handled the remainder of the conversion including the scanner and feed arrangements himself. His diary[5] reveals the many problems he encountered leading up to 7 March 1943, when the complete equipment first worked on the bench at Defford. On 11 March he made the first flight in Stirling N3724.

The results obtained by Beeching in these first flights with a 3 centimetre H₂S were moderate. When flying at 5000 feet he obtained a good PPI picture of the River Severn and located towns such as Cardiff at 22 miles. At 10 000 feet altitude it was a different story. The River Severn could be seen at a range of 16 miles and Worcester, located at 22 miles, could be followed only to 16 miles when it disappeared from the PPI. Evidently much development was needed before an operational 3 centimetre H₂S could be evolved to give good results from an altitude of 20 000 feet.

American Intervention

While we were working on the improvement of this experimental 3 centimetre H_2S system a dispute with the Americans suddenly erupted. As already related (Chapter 16) the Americans had expressed concern about our development of H_2S. However, early in 1943 there was a complete change of attitude. First the evidence of the operational success of H_2S was clear after the initial raids in January and February 1943. Second the American 8th Bomber Command after early experience of operating over European skies began to demand that H_2S be fitted to their bombers (see Chapter 23). The first evidence we had of this *volte face* occurred suddenly in late May 1943. Robinson and a strong contingent of American scientists from the MIT Radiation Laboratory, including the director Lee DuBridge, Bonner and MacGregor, arrived in England with a proposal to replace our 10 centimetre H_2S in the Pathfinder Force with a newly developed 3 centimetre American system (coded H_2X).

> Lovell had immediately contacted me when he and Dee were confronted with the proposal, for it was obvious that in some quarters of the Air Ministry and the Ministry of Aircraft Production the idea found favour. I agreed to go into battle against this scheme. I was bitterly opposed to this suggestion, because I had little faith in the American ability to develop an equipment that would meet our operational requirements. Their whole conception of bombing was entirely different from ours. I knew that a really worthwhile equipment could be developed only as a result of the most intimate liaison between the operators and the Boffins. If the equipment was to be developed in the USA, this liaison would be impossible. I consulted Saundby and received his permission to tell Renwick, Watson Watt and Cherwell that Bomber Command would view such a step with much disfavour and would do all in its power to oppose the transfer of the future development of H_2S for Bomber Command to the Americans.[6]

A major political battle erupted. The trouble was that Cherwell had visited the USA and had been given a demonstration of this American H_2X comfortably fitted in a Boeing passenger plane with a full paraboloid as scanner giving an uninterrupted forward view. Naturally, the results when flying at 5000 feet altitude were quite impressive but bore little relation to the performance of a 3 centimetre equipment flying at 20 000 feet in a cramped bomber.

> June 14 [1943]. We had a heck of a time at work because Robinson and other Americans have come over with a hybrid scheme for a 3 cm H_2S. A ... lot of meetings ending in a CIE [sic. CCE i.e. Renwick] one with Cherwell present, but we won and our 3 cm version (Mk 3) has now got to go with another flap.

The critical meetings in question occurred on 7 June in the Air Ministry. The entries in my TRE record read as follows.

1400 hours   Watson-Watt held a meeting with Dee, Du Bridge, Bonner, Robinson and Lovell. It was *agreed* with the Americans reluctant that the British proposal for converting 200 H_2S Mark II to X-band working should be accepted and that the Americans would concentrate on X-band ASV.
This was merely an advisory meeting, however and we all re-assembled at

1620 hours   in CCE's office with a vast throng including Lord Cherwell, DG of S [Director General of Signals], and the AM [Air Ministry] radar crowd, at which everything was batted through once more, but after much breathtaking suspense it was *agreed* that the British proposal should be accepted and that the target should be *3 Squadrons by Christmas 1943!*

Those acquainted with the many 'crash programme' developments of the wartime TRE will understand the sequence of meetings following this decision. These involved ourselves with the representatives from MAP who had to place the official contracts. Many firms were now involved in the production of the various items of centimetre equipment and nearly all the units required modification in the conversion of the 10 centimetre H_2S to X-band. It is hardly surprising that with all the other pressures on ourselves and these firms the programme went awry so that, of the position in September, as quoted earlier in this chapter, Saward wrote: 'The hope of having some 3-centimetre H_2S in November/December through official sources was, by September, a forlorn hope...'[6]

The Craziest Plan Possible

It was because of this forlorn hope, at the moment in early September when the vision of X-band H_2S appeared as the solution to the Berlin problem, that we hatched the plan as described above by Saward to build six 3 centimetre H_2S systems in TRE for installation in six Lancasters of the Pathfinder Force.

The large wall map of Europe in my office was completely covered with a new kind of chart. In May 1945 I wrote of this in the TRE record:

We hatched a desperate plan to equip 6 of the PFF Lancasters by the end of October with an experimental X-band equipment ... the progress chart which is still on the office wall (May 1945), starts on September 14th 1943 and has jobs for nearly all the H_2S people ranging from responsibility for mounts of the h.f. units to passing out the Lancasters.

On reflection, this appears as the craziest plan possible. The performance of the one X-band system flying in a heavy was so bad that we daren't show it to anyone, and Killip† was so shocked with the performance that he had little interest in flying it. Perhaps the boldness of the plan saved it. We now accumulated a reasonable amount of X-band equipment and soon had more than one aircraft to fly it in. The extra interest and intense hard work all helped and on November 13th the first 3 Lancasters JB 352, 355 and 365 were delivered to PFF. Towards the end a converted Killip‡ was flying night and day and the delivery of the 6 was completed by November 17th.

Flight Lieutenant Len Killip, an ex-operational navigator, had first become acquainted with H₂S during the BDU trials of the original 10 centimetre H₂S in the autumn of 1942. In the summer of 1943 when the Americans demanded that we should fit H₂S in their bombers (Chapter 23) Killip was posted to us (1 August) in order to assist with the test-flying in the Fortresses and Liberators. In September, when the plans for equipping the six Lancasters with 3 centimetre H₂S was made, Saward arranged to continue the posting of Killip to TRE as our Navigational Liaison Officer—a most timely move since I do not believe that the members of my own group could possibly have carried through the development, fitting, and flight-testing of these six Lancasters in the two months after the plan was made.

Our main problem throughout these autumn months had been to devise a scanner that would be free of the gaps and fades that had bedevilled the early 10 centimetre H₂S systems. The problems were greatly exacerbated when we simply scaled the feed arrangements used on 10 centimetres to work with the 3 centimetre equipment and as mentioned above, in Beeching's first flight, even at 10 000 feet altitude, the town of Worcester appeared on the PPI screen at a range of 22 miles but disappeared at a range of 16

---

† Killip had flown in Bomber Command operations in 1941 and was posted in January 1942 as one of the founder members of Flight 1418 (in the summer of 1942 this became BDU—Bombing (or Bomber) Development Unit, see footnote p 106). Originally at Marham and then Tempsford aerodrome, BDU was based on the aerodrome at Gransden Lodge during our association with the Unit. Killip had been one of the primary navigators on the original service trials of H₂S at BDU in the autumn of 1942 and in December he was attached to the 8 Group (Pathfinder) Squadrons to join O'Kane in the training of the navigators in the use of H₂S.

‡ Although this appears in my diary note, when I was in contact with Killip during the writing of this book he responded that he 'never needed converting. The potential of Mk III was evident from the start. What troubled me was the evident deficiency in power. Without ground returns we could not see the lakes in Berlin or elsewhere ... I think the main thing to swing the balance, apart from dedicated work on scanner polar diagram, cleaning up losses in circuitry, waveguides etc., was the availability of the American 3 cm magnetron in the place of the British one, which achieved rather more power, and more consistency. [7]

miles. As described in Chapter 10, in order that the response should be independent of the range the vertical polar diagram of the scanner should follow a cosec^2 law. A crude approximation had been obtained by using deflecting baffles and a shaped waveguide feed on the 10 centimetre H_2S equipment then in service.

A group in the author's office, autumn 1943, planning the fitting of the X-band H_2S in six Lancasters for the Pathfinder Force. The large progress chart is on the facing wall. Reading from left to right of the photograph: the author, F/Lt Killip, F/Officer Richards (standing), Dr F C Thompson, F/Lt Hillman (pointing at the magnetron on the desk), N Z Alcock and F/Lt Ramsay.

We were not lacking advice from the theorists and other aerial experts in TRE about this problem and at the time of the decision to equip six Lancasters with 3 centimetre H_2S we were about to flight-test a new type of scanner. The previous scanners had been sliced paraboloids. That is both the vertical and horizontal shapes were sections of a paraboloid. In this new type the horizontal section remained parabolic to give the best resolution but the vertical section was 'distorted'. The 'distortion' was to make the vertical section through the scanner partly circular and partly parabolic. The top half of the scanner was a section of a barrel while the lower half was parabolic. This arrangement, commonly known as a barrel scanner, was a major improvement on the truncated paraboloid in that the amplitude of the radar responses from a town were far more uniform as the angle of look (that is the range) of the town decreased as the aircraft homed on to it. The profile of the eventual production barrel scanner is shown in figure 22.1. The reflector was a truncated paraboloid, 36 inches diameter × 9 inches × 14 inches wide, welded along its diameter to a 90 degree 'barrel' section (parabolic longitudinal section same as paraboloid, circular cross section with 9 inches maximum radius). It was front fed with a horizontally polarised waveguide horn of 1 inch × 1½ inches aperture. The centre of the horn aperture was displaced 1/8 inch out and 9/32 inch up from the focal point, and the angle of tilt of the horn was 10 degrees into the barrel section. The reflector tilt was − 12 degrees. The rotating mechanism was the same as that used for the Mark II scanner but rotational speeds from 35 to 60 RPM could be obtained by the use of a remote control rheostat in the navigator's compartment.

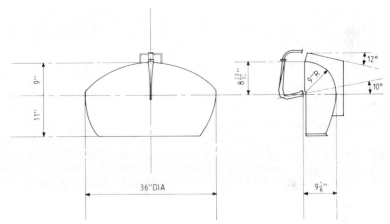

Figure 22.1 Reflector and Horn Feed of the 3 centimetre barrel scanner.

We tested our first crude 3 centimetre barrel scanner in flight at the end of September. I had been away from TRE for a few days and then:

Wed. Sept. 29 [1943] Monday morning I found that the barrel scanner still hadn't flown ... should go off tomorrow. The 6 Lancs are going on OK in all respects except that we don't know what performance we'll get, and all our eggs are on the barrel, so things are pretty fine.
Oct. 11th The 3 cm Lancs are not looking too bad. Every flight seems to get a little better.

In the event, the six 3 centimetre H₂S Lancasters as delivered to the Pathfinder Force had a very good performance. With a peak power output of 15 to 25 kilowatts the average maximum range on towns was about 24 miles when flying at 20 000 feet and the minimum range was less than two miles. Although there was a slight fade in the region of 7 to 13 miles and the returns from seven miles inward were stronger than those in the outer zone, this did not mislead trained navigators and the immense value of the improved definition and coverage was soon manifest operationally.

> <u>Nov.11th</u> [1943] Tremendous pressure on our 6 Lancs. I have a feeling that B.C. [Bomber Command] are about to make an all out effort on Berlin. I told Bennett last night that we'd try to deliver 3 on Sunday and 3 on Tuesday. <u>Nov.21st</u> The 6 Lancs have been delivered, 3 went last Saturday week and one called at Abingdon on the way to give Lord Cherwell a demonstration.†
> All went well thank heavens after a very nervous day ... The other 3 went on Thursday and arrived at Wyton just as PFF were taking off for Berlin! One of the 1st 3 went but unfortunately broke down—a neon in the receiver. This is all terribly exciting. Killip, Beeching and Garner are at Wyton to give them a good start.

The next night the Pathfinder Force used the 3 centimetre H_2S to mark Berlin with great success. Now, squeezed by Bomber and Coastal Commands it was difficult even to find a moment to write a diary note and the next entry is not until—

> <u>Dec. 9</u> [1943] The week before last was terribly exciting since they used the Lancs on Berlin with great success. On the 22nd 2 got there, saw the Templehof and put down their markers within 100 yards of one another. 750 followed up! There were two more major raids on Berlin that week and apparently the most terrific havoc was caused. On the 24th I drove up with Hinckley since there seemed to be some trouble with the servicing and there was a distress call from Beeching. Found incredible inefficiency at Wyton but all the operators most thrilled with the new toy. On the 2nd Dec. there was another Berlin do and 3 out of 4 broke down. This caused a major rumpus from PFF on serviceability but the next night Leipzig appears to have been completely devastated by one raid!

Photographs of the PPI of this 3 centimetre H_2S taken during the Berlin/Leipzig raids are shown in figures 22.2, 22.3 and 22.4.

The devastation resulting from these attacks on Berlin and Leipzig has been described by Saward.[8] He quotes the diary of Goebbels: 'The English aimed so accurately that one might think spies had pointed their way'. Nearly 40 years later when, for me, those events were a mist in the memory, I received a letter that reminded me of the success of those six Lancasters. It was from a resident of the United States, who as a boy, was in Berlin during those raids. 'The realization that the English could see through the clouds caused nothing short of consternation in the Air Ministry, where my father heard the news.'

† Killip was with this Lancaster and flew at only 9000 feet to be certain that Cherwell received a good impression.

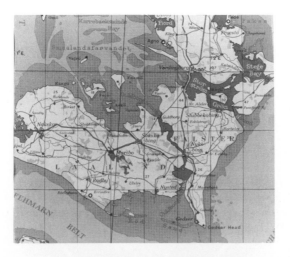

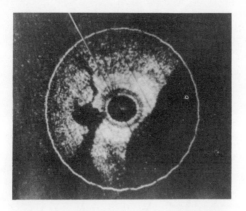

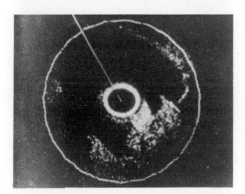

Figure 22.2 Two successive photographs of the PPI in one of the six H₂S 3 centimetre Lancasters taken when the bomber was flying over the Guldborg Sound at an altitude of 21 000 feet on the night of 16–17 December 1943. The straight line on the PPI is the heading marker and the circle is a range marker.

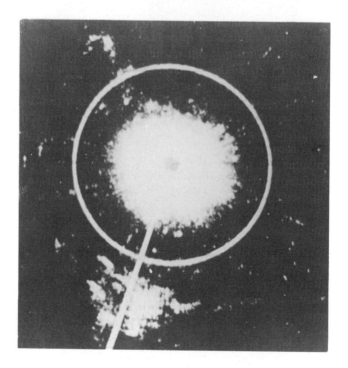

Figure 22.3 The 3 centimetre H$_2$S picture of Leipzig as seen on the cathode-ray tube of a Lancaster bomber as it approached the city from slightly East of North on a bombing raid during the night of 3–4 December 1943. The bomber was at 20 000 feet. The circle is a 10 mile range marker and the straight line is the heading of the bomber.

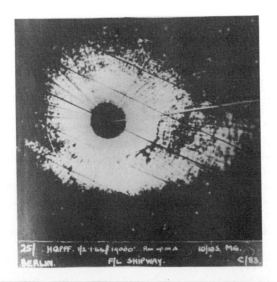

Figure 22.4 The 3 centimetre H_2S picture of Berlin as seen on the cathode-ray tube of a Lancaster bomber of the Pathfinder Force during the run up to bomb release during the night of 1–2 January 1944. The bomber was at 19 000 feet. The straight line running horizontally to the right of the picture is the heading of the aircraft. The black circular area is immediately beneath the bomber and beyond, the line of Tegelersee and Wannsee and other features of the city are visible.

References

1 Webster Sir Charles and Frankland Noble 1961 *The Strategic Air Offensive against Germany 1939–1945* vol 2 (London: HMSO) p 195
2 Webster Sir Charles and Frankland Noble 1961 *The Strategic Air Offensive against Germany 1939–1945* vol 2 (London: HMSO) p 197
3 Saward Dudley 1984 *'Bomber' Harris* (London: Cassell) p 219
4 Bowen E G 1987 *Radar Days* (Bristol: Adam Hilger) p 187 *et seq*
5 Beeching H$_2$S Mark III diary in Lovell TRE record p 21
6 Saward Dudley 1984 *Bernard Lovell* (London: Robert Hale) p 93 *et seq*
7 Correspondence Killip/Lovell April 1990, deposited with the Lovell papers in the John Rylands University Library of Manchester.
8 Saward Dudley 1984 *'Bomber Harris'* (London: Cassell) p 219 *et seq*

Chapter 23

H_2S and the American 8th Bomber Command

The nucleus of the American 8th Bomber Command Headquarters was created in London in February 1942 with a small staff under the command of Brigadier General Ira Eaker. The main body of the 8th Bomber Command followed during the spring and early summer and on 17 August 1942, the Command carried out its first attack against marshalling yards in France. In February 1943 Harris and Eaker had received the directive agreed by the Combined Chiefs of Staff at the Casablanca Conference (see Chapter 21). Whereas the RAF Bomber Command operations were primarily at night, the policy of the US 8th Bomber Command was based on high-level bombing in daylight but they quickly found that the weather conditions in the European theatre were a serious handicap to this policy. An account of the difficulties and of the steps taken by General Eaker to use blind bombing methods with the help of Gee and Oboe has been given by Guerlac.[1] The problem with Gee was that it was essentially a navigational and not a blind bombing aid and with Oboe that only two Liberators and two Fortresses were equipped by March 1943.

By that time General Eaker was receiving enthusiastic reports of the successful use of H_2S by the Pathfinder Force and after discussion with the Chief of Air Staff (Air Chief Marshal Portal) a formal request was made for the installation of H_2S systems in the 8th Bomber Command. On 15 March 1943 Eaker stressed the urgency to Portal:

> I feel very strongly that we should press our plan to have sufficient quantities of this equipment to enable us to effect bombing from above overcast by late summer or early fall, when there can be expected to be a large number of days when high level bombing would be impossible if the target must be seen.[2]

Instantly the task fell to my group. I was at Chivenor with the Coastal Command Wellingtons but was summoned urgently to Defford for a meeting with the Americans on 18 March, only three days after Eaker's letter to Portal. The demand to equip Fortresses and Liberators with H_2S was most unwelcome. We were then involved with equipping the LCTs for Combined Operations as well as with the H₂S in the Pathfinder Force and ASV in Coastal Command. The early doubts of the Americans about H₂S (Chapter 16) were fresh in the memory, but any feelings of satisfaction at this *volte face* were submerged in the imposed strains of these new demands.

On 10 June 1943 Harris and Eaker received a new directive. Although the primary objective of the bomber force was to be that laid down in the Casablanca directive (Chapter 21) the new directive, which came to be known as *Pointblank*, gave more specific directions because of the need to check the growth and reduce the strength of the enemy's day and night fighter forces. The directive assigned more specific targets to be attacked by the RAF Bomber Command at night and by the US 8th Bomber Command in daylight. In particular, 'when precise targets are bombed by the 8th Air Force in daylight, the effort should be complemented and completed by the RAF bombing attacks against the surrounding industrial area at night'.[3]

The first Fortress of the 8th Bomber Command arrived at Defford in June coincidental with the issue of this *Pointblank* directive. Instantly we faced the problem of making our equipment operate at 30 000 feet. Equipment originally designed for operating at a ceiling of 15 000 feet had given endless troubles with high voltage spark-over and breakdown at 20 000 feet. Without any form of pressurisation these problems were greatly increased at 30 000 feet and it was rare for a test flight to be completed without a breakdown of the electronic equipment. But the fundamental problem, once more, was the effect of the increased altitude on the polar diagram of the scanner. The British desire to assist the American 8th Bomber Command to add their weight to the RAF attacks on Germany renewed the intense political pressure and, inevitably, my small team began to crack.

> The real trouble was that the aircraft were required to fly at 30,000 ft. and this presented an entirely new set of difficult problems for the polar diagram. Ramsay and Hillman bore the brunt of this, as well as the screams about the H₂S in the British aircraft. Day after day one or the other flew at 30,000 ft. and took measurements without respite. Eventually, Hillman cracked and ended up in Ronkswood Hospital with pneumonia. The pressure was unrelenting, and on July 23rd CCE [Renwick] convened a meeting, clearly having received a kick from somewhere above, and ordered us to do this work on a priority above all else![4]

At least one good result was the decision to send Killip to us early in August from BDU (see Chapter 22).

> August 14 [1943] The Fortress ... after a long struggle by Ramsay looks reasonably good now. Killip is with us and an enormous help especially with Hillman in hospital.†

† Killip's first flight from Defford was on 4 August 1943 in a B17F Flying Fortress 25793.[5]

There was no American organisation in the 8th Bomber Command based on Alconbury for handling this H_2S. Fortunately amongst my group there was a young Canadian officer, Flt Lt Joe Richards, imbued with a feeling of superiority over his American equivalents. I never discovered how he acquired the large group of American airmen and mechanics who suddenly appeared at Defford. On 23 August I sent a note to Dee with lists of the equipment which Richards had supplied to the American 8th Air Force.

We are carrying the complete baby for the H_2S Fortresses in the 8th Air Force. A constant chase from General Eaker downwards is going on in order to get these Fortresses ready. We are always assured, even by the C/O of the Path Finder Station, that everything at Alconbury is in perfect readiness. In spite of this, when the first Fortress was ready to go, not a single thing was ready at Alconbury—the workshop was not equipped, no bench set was ready and there were no power supplies. Someone somewhere is obviously completely lacking a sense of responsibility ... We have already had to assume responsibility for the Column 9 fitting and flight testing of the four Fortresses and Liberators. In addition we are having to train navigators in order to cover the failure of the arrangements made by the 8th Air Force with Air Commodore Bennett and BDU to do this. But for Richards and our internal organisation, the whole programme would be a complete shambles.[6]

In this way we eventually equipped a total of 18 Fortresses and Liberators for the Americans but not without further tragedy. In November one of the young RAF airmen loaned to my group, Leading Aircraftman Minsen, lost his life through lack of oxygen and extreme cold while flight-testing H_2S in a Fortress at 30 000 feet.

That is the real background to the bland comment in the official American History that, 'The 8th Air Force requested H_2S sets from the British in earnest in March 1943 and the first installation in a B-17 was apparently made during March and April at Defford.[7]

Although a single H_2S-equipped Fortress had dropped two tons of bombs on Frankfurt on 17 August the first major raid of the 8th Air Force guided by H_2S took place on 27 September (1943). On that day over 300 Fortresses attacked the port of Emden. There was thick cloud and the major bombing attack was carried out on marker flares released by the H_2S aircraft. A second attack led by the H_2S equipped Fortresses was made on Emden on 2 October. Subsequent reconnaissance showed that very severe damage had been inflicted on the most vital parts of the target area.

After that attack General Eaker wrote to R A Lovett, the US Assistant Secretary of War for Air, about the gratifying results of the 'first overcast bombing mission ... using the H_2X ...'[8] (H_2X was the American 3 centimetre H_2S equivalent to the British H_2S Mark III). However, it was not H_2X but the British H_2S that guided these American bombing raids. In fact the first bombing mission of the 8th Air Force in which H_2X-equipped planes were used did not occur until 3 November when nine H_2X planes led 60 bombers in an attack on Wilhemshaven.

The account of the development of the American H_2X and its introduction to the 8th Air Force has been given by Guerlac.[9] After the initial raid using H_2X on 3 November the American 3 centimetre-equipped Fortresses 'took the headlines if not the work from the H_2S-equipped Fortress aircraft'.[10] We had honoured the order given to us to equip 18 American bombers with H_2S but with this task completed before the end of 1943 we had no further links with the 8th Air Force. Our association with the 8th Air Force in 1943 is a curious historical episode reflecting very unfavourably on the attitudes of the American scientists and the emigrees to the USA from TRE who, unaware of urgent operational realities in the European theatre, did not believe that H_2S would work until its successes in the hands of the Pathfinder Force were so clearly demonstrated at the end of January 1943. For our part we were thankful to be able to concentrate all our resources once more on the many problems of our H_2S in Bomber and Coastal Commands.

References

1 Guerlac Henry E 1987 *Radar in World War II* vol 8 in *The History of Modern Physics* (Tomash Publications and the American Institute of Physics) Chapter 38
2 See reference 1, p 769
3 See Saward Dudley 1984 *'Bomber' Harris* (London: Cassell) Chapter 18
4 Lovell TRE record
5 From Killip's log book. Correspondence Lovell/Killip April 1990, deposited with the Lovell papers in the John Rylands University Library of Manchester.
6 See reference 4, 23 August 1943
7 See reference 1, p 776
8 See reference 1, minute 75 p 812
9 See reference 1, p 770 *et seq*
10 Lovell TRE record

Chapter 24

The Problems with H₂S in Bomber Command

The diversion of part of our effort to the installation of H_2S in the American bombers during the summer and autumn of 1943 was a serious distraction from our attempts to improve the performance of the H_2S equipment then operating in Bomber Command. The original 'crash programme' had been for the Pathfinder Force. In spite of the many deficiencies of this H_2S Mark I the particular skill of the navigators of the Pathfinder Force, and the special training which it had been possible to give in respect of the use of H_2S, ensured its successful deployment in the early months of 1943.

In May 1943 the main large-scale production of H_2S commenced. Known as H_2S Mark II this was essentially the same as the crash programme H_2S Mark I. There had been certain minor display improvements, rearrangement of some of the controls and facilities for using beacons but these did not affect the fundamental performance of the system. This H_2S Mark II was soon fitted into several of the main force Bomber Command squadrons. In October 1943, for example, H_2S sorties represented one-sixth of Bomber Command's total. Although the serviceability had improved considerably (the H_2S equipment failed in 16 per cent of the sorties compared with 30 per cent failures in July) this use of H_2S by the main bomber force squadrons illuminated the several deficiencies of the equipment of which we were already well aware. A major outstanding need was to improve the definition and the steps taken in respect of the development of a 3 centimetre version have already been described in Chapter 22. Simultaneously we were trying to introduce several other significant improvements and some of the more important will be mentioned in this chapter.

Stabilisation of the Scanner

The scanner was fixed to the fuselage of the bomber and hence the radar beam shared in the pitch and roll of the aircraft. The beam was fan-shaped and narrow in azimuth. Hence, as far as the pitch of the aircraft was concerned this was of no great consequence compared with the serious effect of the roll. In this case violent evasive action caused the radar beam either to dip into the ground or lift away from it. The result on the PPI was therefore either very heavy ground returns or nothing. This considerable extra confusion, introduced at a critical part of the sortie (near the target), proved a very serious deterrent to the effective use of H_2S, and urgent development was undertaken in mid 1943 to roll stabilise the H_2S scanner. The experimental development was not a difficult matter. A suitable motor, servo linked to the angle of roll and driving the scanner framework so that the reflector always rotated about a vertical axis was relatively simple to arrange. At the end of September 1943 we were ready to flight test our experimental model. There were few problems and at a CCE (Renwick) meeting on 5 November 1943 targets were agreed for the conversion of both the 10 centimetre H_2S and 3 centimetre version to roll stabilisation. Three main units were required. The stabilising platform, the modified scanner and the gyro control unit with appropriate electronics. At that time the main manufacturers of the scanners were Nash & Thompson and Metropolitan Vickers. The former firm was bearing the main burden, producing about 80 per cent of the scanners and their promise to make a prototype stabilised scanner (Type 71), see figure 24.1, for use on 3 centimetres by 15 December (1943) seemed to be readily attainable. The target was not met and, as will be summarised later in this chapter, the entire programme went awry together with the other important changes which we were attempting to introduce at that time.

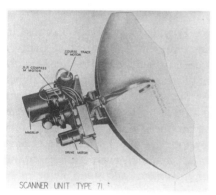 

Figure 24.1   The roll stabilised barrel 3 centimetre scanner (Type 71) and the stabilising platform.

The PPI Display

The most important of these changes concerned the PPI display. The early tests of H_2S in the Halifax V9977 during the spring of 1942 used a plan position indicator with the time base rotating in synchronism with the scanner. The responses on the tube thereby gave the radar picture of the ground area over which the aircraft was flying with the centre of the tube representing the ground immediately beneath the aircraft. Very quickly it became evident that this somewhat crude arrangement needed various changes in order to make the system operationally useful. Cherwell had been at pains to emphasise that H_2S must be a bombing aid and not a navigational aid, but, as soon as the experimental systems were demonstrated to RAF personnel with operational experience, the urgent need of navigational facilities and the potential use of H_2S in this respect became apparent.

Fortunately a major issue became obvious in the earliest flights of the Halifax V9977 in April 1942. On 28 and 29 April, O'Kane was demonstrating the H_2S in the Halifax to Flight Lieutenant E J Dickie (see Chapter 10 p 102). Dickie had been trained as a navigator and as he looked at the map on his table, north was always to the top. But on this Halifax PPI presentation the top of the tube was the direction in which the aircraft was headed since the time base sweep had been lined up to synchronise with the rotation of the scanner. Thus, the north of the PPI display was only at the top of the tube when the aircraft was headed north. O'Kane and Dickie realised during those flights the urgency of devising a system so that the PPI radar picture would always appear with north at the top of the tube irrespective of the heading of the aircraft. All concerned immediately appreciated the force of this argument and in the first EMI 'crash programme' (Mark I) production the phase of the PPI time base with reference to the azimuth of the scanner was controlled by the aircraft's d.r. compass so that the top of the PPI picture always corresponded to true north. This was accomplished via a magslip mounted on the scanner. The rotor of the magslip was geared to the scanner shaft and rotated in synchronism with the scanner. There were two stators on the magslip through which waveforms were generated with the two outputs 90 degrees out of phase. These outputs from the stators were taken to the PPI tube to develop the appropriate time base. In order to maintain constant north at the top of the PPI the difference between the compass mounting and the needle was utilised to drive a repeater motor on the scanner which rotated the stators of the magslip to give the necessary phase correction of the PPI time base.

At the same time a heading marker was introduced on the PPI. This was generated by arranging for a contact to be closed when the scanner was in the dead ahead position. This caused a pulse to be applied to the grid of the PPI for two consecutive rotations of the scanner through the dead ahead position. Thus in the earliest operational H_2S bombers the PPI screen had north at the top of the tube and a bright radial line from the centre to the edge of the tube indicating the aircraft's heading.

Amongst the modifications we attempted to introduce during 1943 was one that enabled this heading marker to be changed to a track marker during the bombing run. The drift angle obtained from the Mark XIV bombsight was used to drive an 'M-motor' which closed another contact on the scanner, offset from the course marker contact by the amount of the drift angle. A switch on the PPI indicator unit enabled the operator to change the bright radial line from 'Course' to 'Track' for the final stage of the bombing sortie.

The Scan Corrected Display

The most important of the changes proposed in 1943 was the attempt to introduce a new 'scan correction indicator' (Type 184). In the PPI presentation of the original H_2S the radial distance from the centre of the tube was linearly related to the range of the target from which the radar echo was received. Compared with the normal map the PPI picture was therefore distorted since the slant range is the same as the true range only at zero height. As the definition and general performance of the H_2S systems improved, increasing reliance was placed on the PPI picture for navigation, particularly beyond the range of Gee, and efforts were made to introduce a new indicator unit with modified electronics so that the PPI presentation was more closely related to the normal map used by the navigator.

The essence of the problem can be seen by reference to figure 24.2. P is the position of the aircraft, at height PO above the ground at point O. If Q is the town, coastline or other feature producing the radar echo on the PPI, then PQ is the slant range as measured and indicated by the echo on the PPI when the time base is linear. For comparison with the normal map and for accuracy in bombing the echo response on the PPI must be related to the ground range OQ. Thus the problem was to devise a means of converting the dimension PQ to OQ before presenting the echo of the object on the PPI. Clearly the amount of correction needed depends on the height of the aircraft above the ground PO. The time base of the PPI must not start until a radar response is received from the point O immediately beneath the aircraft and it must then be non-linear in order to produce the required slant-range correction.

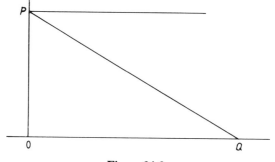

Figure 24.2

In the ideal case the non-linearity of the time base must be hyperbolic with the precise form depending on the height PO of the aircraft above ground, and the range scale in use (in the limiting case where PO is zero the time base is linear). The detailed calculations have been given by Carter[1] and the method of generating the appropriate hyperbolic waveform for the time base was devised by F C Williams.[2]

The indicator units in the Mark I, II and III H₂S equipments housed two cathode-ray tubes. The larger at the top of the unit was the PPI. By setting a scan switch the operator could select three ranges—100 mile, 30 mile and the 10 mile bombing scale. A smaller tube below this PPI served as a 'height tube'. A linear time base ran vertically on the face of this tube and the first echo response was from the ground immediately beneath the aircraft. After appropriate calibration had been made the operator could readily determine the height of the aircraft by setting a pointer to the beginning of this ground echo.

In the ideal scan corrected indicator the appropriate hyperbolic waveforms for the time base would come into action automatically as the height and scan range was changed. This ideal was not approached until later forms of H₂S. The type 184 scan corrected indicator which we were attempting to introduce in 1943 was an approximation in which the appropriate waveforms were generated for the various scan range positions but the operator fed in the height correction manually. A 'distortion corrector' switch mounted to the top left of the PPI was graduated at 5000 foot height intervals and the operator made the appropriate setting on this control after determining the height of the aircraft from observation of the small 'height tube' below the PPI.

Although neither automatic nor absolute in its correction of slant range to ground range this Indicator Type 184 was an immense improvement on the earlier linear time base uncorrected PPI indicators. Unfortunately our attempts to introduce this, simultaneously with the stabilised scanner and track marker facility, met with serious obstacles and to a confused variety of several Marks of H₂S which was not fully resolved until the strategic bombing campaign in Europe was drawing to a close.

The Various H₂S Mark IIs and Mark IIIs and their Time-Scales

At the CCE (Renwick) meeting on 5 November 1943 target dates for the introduction of the stabilised scanner, and the scan corrected indicator were agreed with the manufacturers, the Air Ministry, the MAP and Bomber Command. At that meeting it seemed clear that the manufacture of the new indicator unit (Type 184) would be a simpler and quicker process than that of the stabilised scanner. An internal TRE note of complaint which I wrote in March 1944, four months after that meeting, includes the following extract from the official minutes.[3]

> The introduction of roll stabilization and scan distortion correction were then discussed. G/C Saward (HQBC) said that Bomber Command were anxious for these features to be introduced as soon as possible. It was <u>AGREED</u> that:-
>
> (a) Provision of conversion units for scan distortion correction to be made independently of the introduction of roll stabilization, first priority being for H_2S Mark III 3 cm sets (to be known as H_2S Mark IIIB) in 6 squadrons of the PFF to be followed by conversion of 10 cm equipment (to be known as H_2S Mark IID.)
>
> (b) Provision of conversion units for the introduction of roll stabilization to the Mark IIIB and Mark IID equipments resulting from (a) above to be made as soon as possible.
>
> (c) The production line of Mark IIB equipment to be changed over to Mark IIC as soon as possible in respect of all equipment for installation in Bomber aircraft, and Mark IID to be accepted as an interim measure. [There followed an itemisation of the various actions to be initiated].

If this agreed programme had been carried out the Pathfinder Force would have had Mark IIIA (that is roll stabilised and scan corrected) at the following rates in 1944; 25 in March, 50 in April, 75 in May, 75 in June and 75 in July. That is, the Pathfinder Force should have had 300 operational Mark IIIA aircraft by July 1944. In the case of the Main Force 10 centimetre H_2S, the conversion to Mark IIC (roll stabilised and scan corrected indicator) should have commenced with 75 in August 1944.

About that CCE meeting on 5 November 1943 and the consequences my TRE record reads:

> We were aghast at these delayed dates, but worse was to follow in the months ahead—we had overloaded the firms, peoples' brains and probably ourselves. The delays were appalling—it seemed that the whole country had stopped working ... Matters steadily got worse and worse. The expectation of D–day turned people's minds to the end of the war and the incidence of the flying bomb attacks put a stop to almost any worthwhile progress. The most serious delays occurred on the Stabilising Platform and by the middle of July 1944 only 3 aircraft were fitted to H_2S Mark IIIA! [i.e. roll stabilisation]. The fitting continued at a slow and halting pace and many moons passed before PFF had a significant force of H_2S Mark IIIA. The S band [10 cm] story is even more pathetic. The Gramophone Company managed to achieve a turnover before the end of 1944 and even a few scanners and stabilising platforms were produced. A. V. Roe failed completely to do anything about the changes in the Lancasters and in spite of representations to the Minister this proved a complete bottleneck. In January 1945 instead of having a whole Command equipped with the best roll stabilised S band equipment, Bomber Command were forced to undertake *ad hoc* squadron modifications. Finally since the Lancaster production was decreasing in favour of Lincolns, attempts to get it

in were abandoned. It *was* the basic equipment for the Lincoln but VE day came before many could be used...

By no means all of these delays were the responsibility of the various manufacturers. The summary of items requiring immediate action listed in my internal March 1944 report reveals that DCD [Director of Communications Development] had not even placed the official orders for the stabilised scanners by that time (although they had clearly been instructed to do so at the 5 November 1943 CCE meeting). Unfortunately amongst the urgent actions listed by me (intended only for the information of Dee and Rowe) was to 'shoot DCD' and this reached the individual concerned through an unknown channel. This still further exacerbated the increasing strains amongst all concerned with the project and did nothing to speed the initiatives needed to produce and fit these various units.

On 6 June 1944 the Allied armies landed on the Normandy beaches and the strategic bombing campaign, for which H_2S had been primarily designed, was effectively suspended for several months. The complex discussions between the British and American Staffs about the organisation of the Air Forces in preparation for *Overlord* have been described by Webster and Frankland[4] and by Zuckerman[5]. These discussions culminated late in March 1944 when the American and British Combined Chiefs of Staff directed that the control of all the Air Forces, including the strategic bombers should pass to General Eisenhower. He delegated Sir Arthur Tedder to coordinate the employment of all the British and American bomber forces. For months previously there had been bitter arguments as to whether the bomber forces should disrupt the railways and communications in northern France in preparation for *Overlord*, or as Harris wished, should continue with the attacks on German cities and industrial targets. The new directive[6] issued to Harris on 17 April 1944 was soon to lead Bomber Command on the 'path of a new campaign in which its objects, as also its methods, were different from anything it had previously attempted on a large scale'.[7]

Saward[8] has described the attitude of Harris to this directive; the great success of Bomber Command in a tactical role in support of *Overlord*; and the return to the strategic role in the autumn of 1944†. The new directive issued to Harris on 25 September 1944 effectively restored the whole of Bomber Command to strategic bombing and to the 'operational climax' of October 1944 to May 1945 described in the official history by Webster and Frankland.[9]

During the summer months of 1944 when the major forces of Bomber

† At the Quebec Conference on 14 September 1944 the Combined Chiefs of Staff agreed to the withdrawal of the Strategic Bomber Forces from the direction of Eisenhower and once more Bomber Command was placed under the direction of Portal, the Chief of Air Staff.

Command were employed in a tactical role there was a temporary shift of interest from H_2S to the more precise forms of bombing aids such as Oboe. During this period, in order to clarify the confusion that reigned in many quarters about the state of the various Marks of H_2S, I compiled (30 June) a table (table 1) which clearly reveals the tardy nature of the introduction of the improved forms of H_2S which we had devised in 1943. The problems and delays threw into stark relief the influence of the Prime Minister's directive on the original H_2S.

References

1 Carter C J 1946 *H_2S: an airborne radar navigation and bombing aid* in *J. IEE* **93** pt IIIA 454–456

2 Williams F C, Howell W D and Briggs B H 1946 *Plan position indicator circuits* in *J. IEE* **93** pt IIIA 1219

3 The official minutes of the CCE meeting on 5 November 1943 were in TRE file D1738 Part II. The extract given here is from the Lovell TRE record (insert at p 30)

4 Webster Sir Charles and Frankland Noble 1961 *The Strategic Air Offensive against Germany 1939–1945* vol III (London: HMSO) Chapter XII, 1

5 Zuckerman Solly 1978 *From Apes to Warlords* (London: Hamish Hamilton) ch 12

6 Webster Sir Charles and Frankland Noble 1961 *The Strategic Air Offensive against Germany 1939–1945* vol III (London: HMSO) p 35

7 Webster Sir Charles and Frankland Noble 1961 *The Strategic Air Offensive against Germany 1939–1945* vol III (London: HMSO) p 39

8 Saward Dudley 1984 *'Bomber' Harris* (London: Cassell) ch 21, 22

9 Webster Sir Charles and Frankland Noble 1961 *The Strategic Air Offensive against Germany 1939–1945* vol III (London: HMSO) p 183 *et seq*

Mark	Brief description	Approx. present state of units	Approx. present A/c position	Policy
IIB	Original (S band) H2S + Fishpond	Main Production: Gram.Co/Nash & Thompson. 800/900 per month	(i) Lancaster and Halifax fitted on production line. Between 1500/2000 so far delivered to Column 7. (ii) 8 Group have 7 Mosquito nose installations.	To be superseded by IIC in Autumn 1944 and by IIIA in early 1945
IIC	S band + 184 Indicator (scan corrected) + roll stab. + improved scanner + track line.	Gram.Co. production of IIB begins turnover in Sept.1944. Certain help from RPU to meet initial aircraft programme.	T.I's being done at firms on Lancaster and Halifax. D of Radar target:- Sept.1944 138 Lancaster III + 30 Halifax VI + 2 Lancaster IV Oct. 1944 138 Lancaster III + 40 Halifax VI + 10 Lancaster IV Nov. 1944 130 Lancaster III + 38 Halifax VI + 15 Lancaster IV Dec. 1944 120 Lancaster III + 36 Halifax VI + 22 Lancaster IV Jan. 1945 289 Lancaster III + Halifax VI + 36 Lancaster IV	To be superseded on production line by IIIA early in 1945. (see below)
IIIB	X band + 184 Indicator (scan corrected)	Crash programme. RPU/Nash & Thompson. 220 delivered up to May 14th 1944 since when IIIA's have been delivered.	(i) 150/200 Lancasters fitted at 32 M.U. and so far delivered to Column 9. (ii) 3 S band nose Mosquitos being converted at 8 Group. (iii) Mosquito to X band all-round looking proto. in hand at Defford.	(i) To be superseded immediately by IIIA. (ii) Mosquitos will stay as IIB
IIIA (N.B. When TR3523 supersedes TR3555 this becomes IIID)	X band + 184 Indicator + roll stab. + track line.	(i) Crash programme RPU/Nash & Thompson. Note: RPU delivered 55 conversion units for IIIB to IIIA in May and are now delivering 50 complete sets of IIIA units per month. The IIIA Stab.Platform and Scanner deliveries from Nash & Thompson are behind, however. (About 25 have so far been delivered). (ii) Main programme. Gram.Co/Nash & Thompson superseding IIC/IIB completely beginning early in 1945.	(i) 1 fitted at 32 M.U. Target 40/month at 32 M.U. (Owing to lack of Stab.Platforms and Scanners it is expected that only 20 IIIA's will be fitted in July, the balance being of IIIBs, but in August 40 IIIA's are expected. (ii) T.I's being done at firms on Lancaster, Halifax and Windsor. D or Radar target:- Feb. 13 Squadrons Apl. 31 Squadrons May All production line a/c.	(i) Crash programme for PFF Squadron only. (ii) All heavies as in previous column.

Table 1 Summary of the main H₂S systems, 30 June 1944.

Chapter 25

Fishpond

The delays in producing and fitting the stabilised scanner and scan corrected PPI described in the previous chapter certainly did not happen to another of our ideas which we devised in 1943. With the help of Gee, Oboe, and then H₂S the Bomber Command attacks on Germany during the 1942–1943 winter achieved a new degree of effectiveness—and so did the German defences. By the spring of 1943 the losses of our bombers were becoming a matter of increasing concern. The Germans had given high priority to defence against the night bombing. It was known that large numbers of single-seater fighter planes were flying into the bomber stream under RDF control and making visual attacks on the RAF bombers.

Of course, the problem of giving the bombers a form of radar warning of the approach of a fighter had not been neglected. A simple modification of the original 1½ metre AI system had been fitted to the bombers. This consisted of a pulsed transmitter feeding an aerial system at the tail of the bomber giving a cone of radar coverage. A radar echo from an approaching fighter was converted into 'pips' that could be heard on the bombers intercommunication system. The device, known as *Monica*, had serious disadvantages. The system would only give warning of the presence of another aircraft within the cone of radiation behind the bomber and gave no indication of the position, height or speed of the aircraft. Furthermore, as the density of our bomber streams increased many would be flying in the radiation cone of Monica and the device often 'pipped' so incessantly that it became more of an embarrassment to the bomber crews who often switched it off during operations.

An examination of the bombers which had been attacked by the German night fighters, but had nevertheless succeeded in returning to England, revealed that the attacks by the German fighters were being made from astern and underneath in order to take advantage of the silhouette of the bomber against the night sky. In the first four months of 1943 Bomber Command lost 584 aircraft—over 4 per cent of the sorties and most worrying because it was known that the approach of lighter nights and improved weather would favour the German defences.

I learnt of these troubles from Saward on a Sunday evening (18 April

1943). Those acquainted with the Malvern Hills will know that a walk along the ridge to the cairn is irresistible on a calm and cloudless early spring evening. In those days Saward was frequently with me but although the bomber losses were causing him concern at that time he had no particular reason to discuss them with me. In fact, he had come to Defford and Malvern to look at the progress on the 3 centimetre H_2S which Beeching had first flown in the Stirling bomber a month earlier (Chapter 22). Our walk on the hills was simply an end of the day excursion and it was pure coincidence that, as we reached the cairn, Saward began to tell me about the German fighter attacks on our bomber streams. Alan Hodgkin had recently been diverted from his centimetre AI developments to take charge of the development of a centimetre system known as AGLT (Air Gun Layer Turret).[1] This was to be a radar controlled rear gun turret which would lock on to an approaching aircraft and identify friend from foe. However, in April 1943 this was only in the early stages of development and Saward knew there was no chance that this could help with the immediate problem of the defence of our bombers against the German night fighters. Saward said that was his worry. Indeed he had every reason to be worried. The previous night (April 16–17) Bomber Command had attacked the Skoda armaments factory at Pilsen, also Mannheim, and had laid mines in enemy waters. 39 bombers had been lost —and of the 327 bombers that attacked the Skoda works, 37 had failed to return—an 11.3 per cent loss. Saward exclaimed:

> What on earth are we going to do for a stop-gap? Then I added that H_2S gave us a good picture of the ground below us, and it was a pity it couldn't give us a good picture of the aeroplanes around us.[2]

But, of course, H_2S could do that. I had been on that spot at the cairn with Hensby just after we arrived in Malvern nearly a year earlier and it was he who had remarked about the splendid view to the east and that he believed there was no higher ground until the Urals. Saward had just said that most of the attacks on our bombers were coming from beneath the aircraft. The H_2S echo from the fighter must therefore be within the blank part of the time base before the first ground return. If there was no higher ground over which our bombers were flying then it should be possible to produce a radar echo from a German fighter up to a range of whatever the altitude of the bomber might be above the ground—15 000–20 000 feet in those days.

> We promised Saward to build him a unit to show the potentialities of warning of fighter's approach on H_2S if he would produce a man to do the construction. The whole affair was to be kept quiet to avoid difficulties at Defford on the trials and elsewhere. Sgt. Walker appeared and commenced to build an indicator unit; meanwhile Summerhayes [an American Air Force Officer attached to my group] did some clandestine flights in an ordinary H_2S Mark II with the centre-zero expanded and produced this exciting photograph of a fighter approaching.[3]

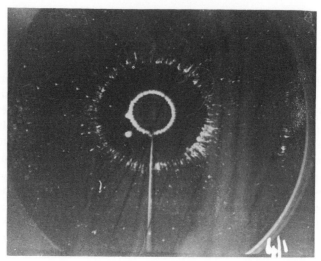

Figure 25.1 The CRT during the Fishpond tests. The approaching aircraft is seen as the bright spot to the left of the line of the flight marker.

Saward[4] has given a detailed description of our private meeting on that Sunday evening and afterwards he acted quickly. Sergeant Walker and two mechanics arrived in TRE the evening after my talk with Saward. The next day he told Renwick, Harris and Saundby what we were proposing. Our task was relatively simple—to make a straightforward change to the time base circuits so that we could display on another indicator tube the initial 20 000 feet of the time base before the first ground returns—hitherto obscured in the central region of the PPI as of no interest.

A few weeks after Saward's visit we installed the new indicator in one of our H₂S Halifaxes (BB360) and on 27 May demonstrated the device to Saward using a Mosquito night fighter to make dummy attacks on the Halifax from astern and below. The radar echo from the fighter appeared as a spot of light on the pool of blackness on the CRT, with its range and bearing precisely indicated. It was a simple but spectacular demonstration and Saundby immediately wrote to the Air Ministry demanding installation in all H₂S aircraft as a matter of high priority. Two days after the demonstration [on 29 May 1943], Renwick summoned an emergency meeting with ourselves, the Air Ministry, MAP and the manufacturers and by the beginning of July the new indicator unit (known as Type 182) was on the factory line.

We had christened the device 'Mousetrap'. This seemed to us a most appropriate code but, as soon as the plans became known outside the initial limited circle, I was handed an urgent signal forbidding us to use this name on the grounds that it would prejudice the security of an impending operation. The alternative code name *Fishpond* was officially allocated and once more Churchill used his power—Churchill to Renwick 9 July: '200 H$_2$S good. Now to round it off 200 Fishponds'.

Compared with our troubles with the stabilised scanner and the scan corrected indicator (Type 184) during that period the results were dramatic. By October 1943 Fishpond was operational and, of the 1030 H$_2$S sorties in November, 553 were equipped with Fishpond. By the spring of 1944 almost every H$_2$S aircraft of the Pathfinder and Main Force carried Fishpond.

The Fishpond unit was known as Type 182. In fact there were two models —an initial crash programme of 200 units known as Type 182 and the main production run differing in only minor details known as Type 182A. As with the main PPI the Fishpond indicator had a radial time base rotating in synchronism with the scanner and with north at the top of the tube as described for the main PPI in Chapter 24. The range from the centre to the edge of the tube was fixed at five miles and, for operating heights at less than this which was the case with the Bomber Command aircraft, the first ground returns appeared as a ring on the Fishpond indicator of a radius proportional to the height of the aircraft above ground. The marker system, which was independent of those on the main H$_2$S PPI consisted of a series of rings on the CRT representing one mile intervals. These could be displayed when required by the operator.[5] Of course, other bombers in the stream appeared as echoes on the Fishpond indicator, but only those rapidly closing in range needed to be regarded as suspect enemy fighters.

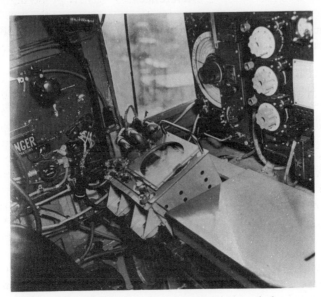

Figure 25.2 The Fishpond indicator unit in the wireless operator's cabin in a Halifax.

The Fishpond indicator unit was normally used by the Wireless Operator and since the display had north at the top of the tube and could also display a heading marker the operator could instantly determine the range and bearing of an approaching fighter. There was no indication of the elevation of the approaching aircraft but experience soon showed that some idea of the elevation of an approaching aircraft could be obtained by banking the bomber. If this manoeuvre caused the echo to cross the heading marker then the suspect aircraft must be well below the bomber. On the other hand if banking caused little displacement of the echo relative to the heading marker then the suspect aircraft must be nearly at the same height as the bomber. Suspicion of a hostile aircraft could also be confirmed by taking evasive action and observing whether the target echo also followed this manoeuvre.

The Fishpond indicator added to the H_2S installation did exactly what it was intended to do; namely to display the radar echoes from any aircraft within a range of five miles of the bomber in the lower hemisphere and, in doing so, enabled many aircraft of Bomber Command to evade attack by the German fighter aircraft. It was never intended as a sophisticated system such as AGLT but this equipment did not become available until the end of the European War. As with H_2S itself, Fishpond met an urgent operational requirement in a remarkably short time-scale but, as will be seen (Chapter 30), like H_2S itself, Fishpond generated its own antagonisms and critics.

References

1   Hodgkin Sir Alan *Chance and Design: Reminiscences of Science in Peace and War* (Cambridge: Cambridge University Press) to be published
2   Saward Dudley 1984 *Bernard Lovell* (London: Robert Hale) p 87
3   Lovell TRE record p 19
4   Saward Dudley 1959 *The Bomber's Eye* (London: Cassell) Chapter 23
5   See Hinckley H G 1945 *Fishpond* TRE Jr. 5 pp 65–73 for the detailed circuits and arrangements

Chapter 26

Conflict with Bomber Command

Of the 53 000 sorties made by Bomber Command in 1943 over 32 000 were led by H_2S. In fact the Command was depending almost entirely on the radar aids initiated in TRE since most of the remaining sorties in that year (20 000) were led by Oboe. Notwithstanding the delays and difficulties we began to experience in 1944 in the attempts to introduce improved features to H_2S (described in Chapter 24), the fact was that a substantial force of Bomber Command aircraft were operating with H_2S. By April (1944) 23 squadrons were being fitted. The Pathfinder Force was operating with 59 H_2S Mark III-equipped aircraft and 118 with Mark IIB. In total, Bomber Command had 638 bombers fitted either with 10 centimetre or 3 centimetre H_2S—365 Lancasters, 214 Halifaxes, 53 Stirlings and 6 Mosquitoes. We should have been reasonably content in TRE, but this was far from the case. There were two major reasons for our turmoil. One was internal arising from arguments about how we should proceed to make an H_2S superior to the various Marks of 10 centimetre (Mark II) and 3 centimetre (Mark III) then in service or with the pending modifications of roll stabilisation and scan correction. These futuristic H_2S systems as then envisaged and developed will be described in Chapter 27. The other reason, the subject of this chapter, lay with the Pathfinder Force and Bomber Command and in the spring of 1944 an acrimonious dispute developed which quickly culminated in a confrontation at the highest level.

The dispute concerned the operational use of H_2S and now, nearly 50 years after the event, many will feel that it was none of our business. The fact that, in 1944, we felt it to be very much our business is simply a reflection of the extraordinary close integration we had with military operations. Indeed, it was this integration that so sharply distinguished the Allies from the enemy, epitomised in the use of H_2S equipment as ASV in Coastal Command (Chapter 18) and in our fitting of the six Lancasters with 3 centimetre H_2S late in 1943. As described in Chapter 22 this special equipment led the devastating raids on Berlin and Leipzig in December 1943 and

January 1944. The first Pathfinder Force Lancasters fitted with the 200 crash programme production of 3 centimetre H_2S began to be delivered on 23 December 1943 and with the rapidly increasing numbers available to the Pathfinder Force an escalation of the successful bombing campaign deep in Germany was to be anticipated.

This did not happen. After their great success on Berlin and Leipzig, Bomber Command made a whole series of ineffective raids and suffered heavy losses. Repeated attacks on Berlin, for example, did not materially increase the damage already done and, as we had so frequently found in the past, poor performances by the Command were readily attributed to the deficiencies of H_2S which suffered acrimonious treatment at the hands of its opponents. However, it soon became evident to us that the fault did not lie with H_2S, but with the use being made of it by the Pathfinder Force.

> Wed. March 22 [1944]—Depressing story from ORS [Operational Research Section of Bomber Command] on Sunday morning. PFF are obviously making many boobs in marking with H_2S.

By this time Killip was by far the most experienced H_2S operator in the RAF. In fact, there was always a danger in asking him to visit an operational base since he was liable to be pressed into that night's operation.[1] But now we were deeply suspicious about the state of the 3 centimetre equipment in the Pathfinder Force and Killip journeyed to 8 Group to investigate. His immediate report was worse than we could have imagined.

> ... the precious 3 cm equipments were being scandalously treated in 8 Group. Instead of being given preferential treatment with especially skilled crew and priority marking as a special force, they were simply being put in as wastage rate aircraft and carried no more weight in marking than an ordinary S-band equipment. The crews trusted with the marking were not specially picked and trained and the general internal happenings in 8 Group appeared to be near chaos.[2]

That in itself would have been sufficient for us to act with vigour but, by chance, a different, although related, issue emerged simultaneously about the use of H_2S. The system had been designed as a blind bombing device. Initially, early in 1943 there were so few bombers equipped with H_2S that the technique of bombing on markers released by a small number of Pathfinder Force H_2S-equipped bombers was essential. Now, a year later, large numbers of bombers in the main force were equipped with the navigators trained in the use of H_2S and it seemed to us that a complete raid of blind bombing on H_2S should be given a trial. In spite of agitation from TRE this trial had never been made and, when the inefficiency of the marking by the Pathfinder Force came to light, we made a vigorous complaint during the course of one of Rowe's Sunday Soviets. From my TRE record:

> Sunday March 26th 1944 ... 11.30 CCE [Renwick] VCCE [Watson-Watt] and DG of S [Tait—Director General of Signals] in Chief Superintendent's office. Had two terrific goes at them, one before lunch on the mix-up in London on such items as the Type 26 stabilising platform. Goodier produced vast charts showing that the bottle-neck was in D of Radar and DGE Department. After lunch with Killip on the mess which PFF were making of using H_2S. This seemed to go down extremely well, especially as we were able to flourish the beautiful [radar] photograph of Frankfurt which had been taken with the 184 [scan corrected] Indicator. As a result of this we were asked to produce a document setting out TRE's views on what should be done in PFF and Bomber Command to remedy the situation.

The issue quickly escalated because of the coincidence that four days later on my way to a Renwick meeting in London I called at Bomber Command HQ to see Saward. In agitation he handed me two letters just placed on his desk. One was a pleasant letter from Renwick to Harris written after the previous Sunday's meeting in Rowe's office suggesting that he should give TRE's views on blind bombing with H_2S a chance. The other was Harris' two page irritable response, beginning: 'Tell TRE to mind its own ruddy business' and continuing with a harangue against scientists and TRE in particular '... they remind me of pimply prima donnas struggling to get into the limelight...'.

This, from the C-in-C of a Command that would have been dead long ago without the scientific aids originating in TRE was too much. When I reported to Rowe I had never seen him so angry. 'Well, if he wants a battle he can have it', and the campaign to 'get rid of Harris' had begun. Rowe consulted Renwick who asked for a suitable paper which he would hand to the Chief of Air Staff. On 3 April the short paper *H_2S in Bomber Command —some TRE views*, signed by Rowe, was handed to Renwick.[3] The memorandum was brief, mainly critical of the accuracy of the recent blind marking using H_2S, and stating that if blind marking was the policy then it must be carried out by specially trained crews and that if blind bombing by the main force was no longer intended, 'then the large scale production of H_2S involved in fitting every bomber, should be carefully reviewed. The effort in the country and at TRE might be much better used in devising and fitting improved equipments in the blind marking force'.

Outwardly, at least, peace reigned for over a week. Then, late on the evening of Friday 14 April Rowe phoned: 'the first broadside has been fired'. He had been in London and as he told me the following day:

> He was very pleased with his visit to London yesterday and had spoken to DB Ops and also ACAS Ops. (AVM Coryton) whom Dobson† told us was

† The Chairman of A V Roe with whom we had recently been in touch over the fitting of H_2S in the Lancasters.

the man sacked by Harris. They were obviously spoiling for a fight along the lines of our document and [Rowe] said that Coryton showed him a minute written by CAS on the document which was very fair.[4]

A week later we were in the Air Council Room, Whitehall. The day was a Saturday (22 April 1944) and I had travelled to London with Rowe and Robinson.† We were early and, oddly, my most vivid memory of that day is of the sunshine and the beauty of the spring day as we paused on a seat in St James's Park. The pelicans on the lake seemed at peace and, in a spirit of truancy which so precisely reflected the atmosphere, Rowe reminded us of such a day in England when we would not be at war. 'Tell you what, let's not go'. But this Top Secret meeting had been summoned for 3 pm and soon sixteen of us were around the Council Table in the Air Ministry.

If my most vivid memory is of the sunshine on that spring day my TRE record[5] reminds me that far greater issues were at stake.

Saturday, April 22nd 1944 ... the critical 'get rid of Harris' meeting took place at 1500 hours in the Air Council Room in Air Ministry. Most unusual preparations had taken place in TRE for this and the Chief Superintendent [Rowe] had done considerable lobbying. The final rehearsals took place between Chief Superintendent, Robinson and Lovell during the journey and by the lake in St. James's Park at 14.15 when we met Matthews‡ for last minute scandal from PFF. The meeting was incongruously placed in the middle of an unusually beautiful spring day... [at] 1500 hours the following assembly materialised

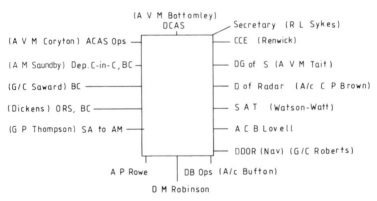

We had certain hopes that CAS himself would take the Chair and that Harris would be made to attend. Chief Superintendent described Bottomley to Robinson as a man who looked at civilians with a face which resembled

----

† D M Robinson had taken over from Dee, who was away ill, as Superintendent 3 (see Chapter 27).
‡ L H Matthews a member of Ratcliffe's Post Design Services group assisting with the introduction of radar into service use. He was a distinguished zoologist (see *Biog. Mem. R. Soc.* **33** 413, 1987).

a fish on a slab on a hot day! On this occasion however Bottomley extended a pleasant welcome and it was soon evident that the battle was already half won before the meeting began. The agenda consisted of a good write up of H_2S with several appendices containing most of our original paragraphs, but embellished and shifted around...[6]

After nearly three hours we emerged from the Air Council Room. My TRE record of the meeting pays tribute to Saundby 'in the difficult position of not appearing disloyal to his Command whilst obviously agreeing with the TRE points'. The major results were as follows:

1 Saundby's confession that only Harris had stopped the use of blind bombing on H_2S by the main force and the pressure from DCAS to do so at the earliest opportunity.
2 Saundby's statement that his ideal arrangement would be for every plane to blind bomb on H_2S and the consequent easement of the planning of operations.
3 The agreement to supersede H_2S Mark IIC by IIIA in *all* bombers as soon as possible.
4 After an attack on the use of H_2S by the PFF, although there was denial of its inefficient use, nevertheless two of Bennett's PFF squadrons had been handed over to 5 Group and the remaining Mark III H_2S in the PFF were being organised into coherent squadrons.
5 Saundby's statement that H_2S Mark III had enabled them to fight the Battle of Berlin and that they had been 'immeasurably effective'.
6 The decision to investigate the fitting of all-round looking H_2S Mark III in Mosquito aircraft because of their low loss rate. †

Rowe had been greatly embittered by Harris's response to Renwick and was disconsolate on the journey back to Malvern that evening that the meeting had not achieved his hope of dismissing Harris. He had no reason to be since every point of technical and operational significance that we had raised had been conceded. In fact, towards the end of the meeting DCAS turned to Saundby with the remark: 'That's another point then on which you have already met TRE's requirements'. Before our journey ended that night Rowe conceded that we had 'undoubtedly proceeded a considerable way'.

It is curious that Bennett, who was so vital a factor in the original introduction and use of H_2S was now suffering our catechism for its inefficient use in the Pathfinder Force. Perhaps the most important result of our intervention was the decision to hand over to 5 Group two of Bennett's 8 Group H_2S squadrons. In that way we became associated with Air Vice Marshal The Honourable Ralph Cochrane the Commander of 5 Group who

---

† At the date of that meeting six Mosquitoes of the PFF were equipped with H_2S but it was a 'nose' installation and therefore not all-round looking.

rapidly organised the two H_2S squadrons into a precision Pathfinder Force. In 8 Group Bennett allowed two of his staff, Whitbread and Barton, to reorganise the whole of his H_2S training and the use of it by the Pathfinder Force.

The result of this conflict with Bomber Command was soon evident in the steady improvement in bombing accuracies during the summer months of 1944. For example, on the nights of 24–25, 25–26 and 28–29 July (1944) Stuttgart—a 'difficult target' for H_2S—was bombed through ten-tenths cloud and nearly 400 acres of the central area of the city was completely devastated.[7] By the autumn of 1944, 50 per cent zones of 1 to 1½ miles were regularly achieved and Cochrane's 5 Group H_2S squadrons were marking with errors of less than 100 yards.

References

1 Personal communication Killip/Lovell 24 April 1990 deposited with the Lovell papers in the John Rylands University Library of Manchester.
2 Lovell, TRE record p 35
3 A copy of this paper marked *Most Secret* but otherwise unidentified is inserted in Lovell TRE record p 35
4 Lovell TRE record p 36
5 Lovell TRE record p 36 and insert
6 The Agenda and notes *Policy in regard to future use of H_2S* with Appendices A to D is marked *Top Secret* but not otherwise identified. Copies are inserted in Lovell TRE Record at p 36
7 For the details of attacks on Stuttgart and other targets in the second half of July 1944 see Webster Sir Charles and Frankland Noble 1961 *The Strategic Air Offensive against Germany 1939–1945* vol III (London: HMSO) p 173 *et seq*

Chapter 27

The New Versions of H₂S

In Chapter 26 I have described our conflict with Bomber Command over the operational use of H₂S. The second reason for our turmoil in 1944 was largely internal in TRE and arose from disagreements about the development of an improved H₂S with superior definition to the Mark II and III systems then in service or projected. Before 1944 ended we had no less than five Lancasters at Defford equipped with various experimental versions of improved 3 centimetre (X-band) or 1.25 centimetre (K-band) H₂S and a sixth system was under development.

(i) Mark IV H₂S

The least disputable of these systems was known as Mark IV H₂S and, unlike the other complex of developments, was a natural successor to the 10 centimetre and 3 centimetre H₂S described in earlier chapters. Mark IV was, in fact, the first H₂S system to be designed as such as distinct from the *ad hoc* adaptations and changes to the various versions of Mark II and Mark III carried out to meet urgent operational needs.

The idea of such a purpose-designed H₂S first arose at the end of 1942, even before the first operational use of H₂S. By that time the research team at EMI had completed their development work and at a meeting on 29 December 1942 we agreed that they would turn their attention to a new H₂S incorporating radically new features. Early in 1943 we defined the system more precisely as a 3 centimetre roll stabilised scanner, automatically scan corrected indicator, bombing circle and drift line on the PPI derived from the Mark XIV bombsight and picture stabilisation. In fact, as described in Chapter 24, several of these features—the roll stabilisation, scan corrected indicator and drift line—were eventually introduced in 1944 as *ad hoc* modifications to the existing 10 centimetre and 3 centimetre systems. The automatic scan correction with altitude, the picture stabilisation, and the bombing markers linked to the Mark XIV bombsight were the essentially new features that did not materialise until this Mark IV H₂S was developed.

In February 1943 we allocated two members of my group, J B Smith and

217

Platt, to work full-time on this new H_2S in collaboration with the research group at EMI. In August and September the initial flight trials of an experimental system were in progress. Although the Mark IV H_2S was conceived as a 3 centimetre system, there was no spare equipment on this wavelength at that time (the available units were in use on the development of the basic H_2S Mark III—see Chapter 22) and these initial flight trials used the 9 centimetre (S-band) transmitter and receiver. By the summer of 1944 an engineered prototype system working on 3 centimetres was undergoing flight trials in a Lancaster at Defford. The new features have been described by J B Smith.[1]

RF Equipment and Scanner

The Mark IV H_2S used the standard radio frequency units that were then becoming available for use in the centimetre wavelength region. The RF unit (TR 3523) used a magnetron working on 3 centimetres producing 35 kilowatts of peak power. The pulse width was reduced to 0.5 microseconds (μs) compared with the 1 μs of the earlier Marks of H_2S and the pulse recurrence rate was increased to 1000 per second. This gave a scan time of 800 μs on the display corresponding to a range of 80 miles. Another refinement was the automatic frequency control of the local oscillator. The scanner was similar to that used in Mark IIIA, that is a barrel type roll stabilised over ± 30 degrees. The rotation speed was variable within the limits of 30 to 70 RPM. There were two differences from the Mark IIIA scanner. The magslip resolver was replaced by a sine potentiometer—a unit that provided direct current voltages proportional to sine and cosine of the azimuth angle of the beam—and these voltages produced the waveforms for the PPI. Although the barrel section of the scanner had greatly improved the coverage (see Chapter 22) a still further refinement was introduced into the Mark IV scanner—a tilt control which enabled the operator to obtain the clearest PPI picture during the final phase of the approach to the target.

Scan Correction and Bombing Circle

The need for scan correction, that is the presentation of ground range instead of slant-range on the PPI has been described in Chapter 24. Although the scan corrected Type 184 indicator introduced into the Mark III H_2S used the hyperbolic waveforms to correct slant to ground range, the altitude had to be set in manually. The significant improvement in the Mark IV display was that the appropriate hyperbolic waveform was produced automatically as the altitude of the aircraft changed. In the scan corrected Mark III system the operator determined the height by setting a marker on the first ground return on the height tube. In the Mark IV this was done automatically by means of a drifting strobe which locked on to the leading edge of the first ground return. The technique by which the strobe was generated and then produced the hyperbolic waveform appropriate to the altitude of the aircraft has been described by Smith[2] and by Carter.[3]

Another important feature of the Mark IV was the integration of the PPI display with the Mark XIV bombsight to produce a bomb marker on the CRT. The linkage of the PPI to the d.r. compass to produce constant north at the top of the tube and the production of a radial line on the PPI as a course marker were features of the earliest H₂S system. Later, by controlling a contact on the scanner by an M-motor driven from the Mark XIV bombsight a radial track line was introduced on the PPI (see Chapter 24). In the Mark IV H₂S the bomb release point was presented on the PPI as the intersection of the track marker with a circle in the centre of the PPI of radius corresponding to the forward throw of the bombs in use. This marker on the PPI was derived from the bombing angle shaft of the Mark XIV computer which was used to drive a potentiometer with a secant law calibration. The combination of this voltage and one representing the height caused a marker to appear on the PPI at a distance from the centre corresponding to the ground range of the target. The intersection of this marker with the track marker would be the point of impact of a bomb released at that instant. The details of the arrangement have been given by Smith[4] and by Carter.[5]

Display Stabilisation

An entirely new feature of the Mark IV H₂S was stabilisation of the PPI display. In all the previous H₂S systems the radar picture of the ground moved over the CRT in accordance with the ground speed and track of the aircraft. The start of the scan was stationary at the centre of the CRT and represented the aircraft's position with the track marker originating from it and the bombing circle around this central point. In the stabilised display, shift voltages are applied to the CRT to correct for the ground miles (north–south and east–west) travelled by the aircraft from a chosen reference time. The radar picture of the ground then remains stationary on the PPI display while the bombing circle and track marker move over it.

The shift voltages necessary to 'freeze' the display were taken from the air position indicator (API) and the air mileage unit (AMU). In the absence of wind the voltages so derived could be appropriately applied to the display to give a complete 'freeze' of the picture. The method of generating these shift voltages and the additional ones necessary to allow for wind has been described by Smith[4] and by Carter.[5]

There were a number of important advantages of this stabilised display. First the actual wind speed at the height of the aircraft could be determined since the wind shift voltage derived from the bombers' AMU and API would not, in general, correspond precisely with the actual wind. The PPI picture would not therefore be completely frozen but would drift across the face of the CRT. By measuring the speed and direction of this drift the divergence of the wind vector from that set into the aircraft's AMU and API could be determined. The PPI display could then be rigidly frozen with the radar picture of the ground stationary on the tube face and the track and bombing circle markers moving over it. This was of advantage to accurate bombing since the identified part of the target could be assessed at ranges of up to 10 miles and not at shorter ranges corresponding to the point of bomb release when the H_2S picture from large built up areas was more difficult to interpret. That is, the aiming point could be assessed at a range where the H_2S picture was clearest, an appropriate mark placed on the tube face and bomb release made when the intersection of the bombing circle and track marker coincided with that mark. In practice, manual shifts were applied to bring the selected aiming point to the centre of a perspex face covering the CRT marked by a small cross. The stabilised display used in this manner during a bombing run gave considerable tactical freedom as regards evasive action since, within reason three points remained defined on the tube, namely the aircraft's position, the bomb release point and the aiming point.

The flight trials of this system had proceeded without difficulty, the engineering development had been completed by the Gramophone Company† to the extent that we anticipated that Mark IV H_2S could be introduced to operational use during the summer of 1944. However, this was not to be because of a number of factors. The *ad hoc* introduction of roll stabilisation, scan correction and track marker to the H_2S Mark II and III had provided these systems with the essential features of Mark IV, except the picture stabilisation, and this had understandably reduced the enthusiasm for the introduction of Mark IV. Furthermore, the Gramophone Company, beginning to produce large quantities of Mark II and Mark III, was reluctant to interfere with this production line and were slow in manufacturing the production prototypes of Mark IV. As will be seen, Mark IV H_2S, although the natural successor to the various versions of Mark II and III, had a turbulent passage in conflict with other versions of high-definition X-band and K-band H_2S systems. Indeed, its introduction to Bomber Command was still the subject of agitated discussion as the Allied armies began their final surge across Europe in the early months of 1945.

(ii) K-band (1¼ centimetres) Versus X-band (3 centimetres) H_2S

Although the Mark IV H_2S was a more sophisticated blind navigation and bombing equipment than any of the existing or proposed 10 centimetre or

† For the relation between EMI and The Gramophone Company see p 134 Chapter 14.

3 centimetre systems, it used the same sized scanner as Mark III and so there was no improvement in the definition of the PPI picture. During 1943 the techniques for radar systems on a wavelength of 1¼ centimetres (K-band) emerged. For the same size of scanner this would improve the definition of the radar picture by more than a factor of two compared with the Mark III (3 centimetres). When such a reduction in wavelength was first discussed in 1943 very little work had been done in the UK on this wavelength. The British effort had been almost wholly concentrated on dealing with the urgent operational requirements for radar systems on 10 centimetres and 3 centimetres. In America, however, where the scientific effort was not so monopolised in this way considerable progress had been made in developing magnetrons and associated TR and receiving equipment for K-band.

Dee, who had been promoted to Superintendent level, now had control of the main centimetric research group as well as the various applications groups amongst which was my H₂S group. Towards the end of 1943 he decided to place Sam Devons† in charge of a party to develop a very high definition H₂S on K-band using the American magnetrons and RF equipment. Devons and his group worked in another of the College buildings where the fundamental research was concentrated and had hitherto had no association with our H₂S. The idea was to fly this experimental K-band H₂S in a Wellington aircraft and Dee, quite reasonably, attempted to establish a liaison between the Devons party and my group in order to assist them with the building of the units, installation in the aircraft and flight tests at Defford, where we now had a substantial resident section of my group.

Dee had overlooked the consequences of the highly competitive spirit between the various groups that had grown in TRE. My own group considered that anything to do with H₂S was our affair and as soon as this plan became known there was a most violent reaction. This reaction materialised in a scheme to achieve the same definition as the proposed K-band by doubling (to six feet) the size of the scanner used on the 3 centimetre Mark III H₂S. We argued (correctly at that time) that the power and sensitivity of the X-band equipment was far superior to that of the proposed K-band and hence the six foot X-band would be a far better H₂S of high definition and great sensitivity.

Before 1943 closed these battle lines had been formed and in the early months of 1944 the only serious dispute of the war occurred between Dee and myself. By mid-January we had excited Bennett about the prospect to such an extent that he produced a sketch showing how a cupola to carry a six foot scanner could be placed underneath the Lancaster and, for his part, Dee had talked to Renwick and persuaded him to convene a meeting to settle the future policy for H₂S.

---

† Devons had joined TRE in the late summer of 1941 and had hitherto been concerned with fundamental centimetre techniques.

My TRE record[6] reads:

> ... the background to one of the most acrimonious H_2S meetings which ever took place. On Friday February 4th 1944 Dee, Lewis, Devons and Lovell journeyed to London to meet Renwick, Watson-Watt, Bennett, Tait, Theak, Saward, D of Radar, etc. The meeting was convened to discuss how, if at all, K-band should be fitted into the H_2S programme, e.g. there had been serious suggestions that a double K/S system [that is 1¼/10 cm] should be devised, using K for bombing and ordinary S [10 cm] for navigation. However, the meeting, following the lead of Bennett, jumped for the 6 ft X which it christened 'Whirligig', said that K-band should 'continue' and more or less dismissed the subject of Mark IV. Dee, in particular, was very annoyed with this and afterwards accused Lovell of fixing the meeting with Bennett beforehand.

Renwick promised to convene a more definitive meeting, but before that took place on 22 February (1944), Dee, to my great dismay, suddenly disappeared from the scene. Ever since joining TRE in the spring of 1940 Dee had been subject to bouts of pneumonia and, on the Sunday [13 February] following the acrimonious meetings, we were told that arrangements had been made for him to take three months leave and that Denis Robinson would act as Superintendent 3 in his place. Dee sensed my perturbation and the day after this announcement we made our last official journey of that era together. Fortunately I had no difficulty in responding to his plea that I should be a 'good boy' since, whatever doubts Robinson may have entertained about H_2S in the early stages he was now an enthusiastic supporter. Within days I was in league with Robinson in devising a typically British compromise to the K/X-band disputes. Ironically that last journey I made with Dee on 14 February (1944) was to A V Roe in Manchester to see Roy Chadwick the designer of the Lancaster. The purpose of the visit was to persuade Chadwick to put a 6 foot blister under the Lancaster to take our 6 foot X-band scanner. It is hardly surprising that we had a 'frigid reception' and we were appalled at how little Chadwick knew about H_2S, although all his bombers coming off the production line were equipped with it. A few days later (on Good Friday) Dobson [Chairman of A V Roe], an enlightened man, came to TRE and subsequently, although we had a multitude of difficulties with A V Roe, we did not believe that these were the fault of Dobson or Chadwick.

After that journey Dee retired from the scene and it was with Robinson that I now faced the decisive Renwick meeting on 22 February to deal with H_2S policy. Devons was also present but the three of us from TRE were faced by the 17 RAF officers concerned with the production and operational use of the equipment. Lord Cherwell's presence added significance to the crucial nature of the decision that had to be made about the future policy. The proposals for a K-band H_2S had become more plausible since 60 magnetrons and RF heads had been promised from American sources for delivery in 1944. Robinson referred to the increase in sensitivity achieved during the last three months and thought that ranges of 20 miles could be achieved. Watson-Watt raised the crucial question of the possible absorption in cloud and our inability to answer this question decisively is indicative of the scientific knowledge existing at that time about these very high frequency radiations.† The minutes of that meeting, so important, as it transpired, to the future of H_2S in the second world war are in my TRE record.[8] As far as we were concerned in TRE we now thought that a limited crash programme on K-band was not only feasible but desirable. Under Robinson's influence we had gone to the meeting with three proposals (i) fit H_2S Mark IV in all Lancasters, (ii) *either* a limited Whirligig (6 foot X-band) *or* a K-band crash programme before the end of 1944, and (iii) to be followed by a general X-band programme.

However, we had discounted the enthusiasm of Bennett and Bomber Command for the promise of immediate high-definition, long range H_2S in the shape of the X-band Whirligig and we came away from Whitehall that evening with the very burden of a commitment to Whirligig *and* K-band we had hoped to avoid.

The comments in my TRE record on that meeting epitomise our reaction.[9]

> After this [that is the K versus X discussion] there was a most depressing discussion on the bulk H_2S production for 1945. The whole attitude was quite different from what it was a year ago. The tendency now is to let production dates dictate policy. Instead of insisting that there must be enough Mark IV H_2S's to cover the Mark IV Lancaster production Renwick and the meeting generally were saying that they must wait to hear DRP's [Director of Radio Production] dates and then decide when to change over the Lancaster IV line meanwhile putting in H_2S Mark IIC. One must always remember this meeting as the meeting which seriously discussed the possibility of getting X–band definition by using a 6ft S-band array because of the difficulty of producing X–band magnetrons...

That was not the end of the policy meetings but in the meantime we settled down to make a flyable version of both the K-band and 6 foot X-band versions of H_2S.

---

† This K-band wavelength of 1.25 centimetres is within the broad water vapour absorption line which peaks at 1.35 centimetres. The detailed measurements were not made and published until after the war but it is possible that Watson-Watt was aware of some of the preliminary indications from the work at GEC Research Laboratories Wembley and elsewhere that there was a possibility of serious absorption by water vapour at this wavelength. For details of the immediate post-war work on absorption of microwaves see B R Bean and E J Dutton.[7]

(iii) **Lion Tamer—K-Band H₂S Mark VI**

After the 22 February policy meeting at which it was decided to proceed with a K-band H_2S the proposed operational system was designated H_2S Mark VI. Lion Tamer was, and remained, the code for the K-band development programme. After that meeting it was agreed that Devons should go to the United States to provide the urgently needed liaison on the procurement of the K-band magnetrons and other specialised components for that wavelength. His team, hitherto concerned with the laboratory development, was split, part to help with the experimental Lion Tamer flights of K-band in the Wellington and part to help with the K-band (Mark VI H_2S) for the operational Lancasters. The first Lion Tamer test flight of K-band in the Wellington occurred from Defford on 8 May (1944), but most of the interest and pressure centred on the prototype Mark VI installation in the Lancaster. This Lancaster (ND 823) equipped with the prototype H_2S Mark VI K-band first became airborne from Defford on 25 June (1944). At least the system worked and the River Severn could be seen on the PPI at a range of two to three miles from an altitude of 5000 feet.

Meanwhile, shortly after Devons' return from his visit to the USA, Renwick summoned yet another policy meeting, such was the confusion and argument prevalent about the many versions of H_2S and the uncertainty about the possible performance of K-band. With the exception of Lord Cherwell, substantially the same group of people assembled in the Air Ministry on 16 June (1944).[10] Devons explained the organisation of the K-band development in the USA and reported that the K-band magnetron was being developed by the Columbia Radiation Laboratories and that two or three from that source should arrive in the UK before the end of June. At that time MAP had requested the British Commission in the US to procure 500 K-band RF heads and it was agreed that a signal be sent to confirm that we did not propose to develop K-band RF heads in the UK and were relying entirely on American production, first on 200 from Sylvania and later, 300 from Philco.

Flight Lieutenant Day, a member of Saward's staff, had recently flown with the Lion Tamer equipment in the Wellington and said he obtained ranges of five miles from 5000 feet. Devons had witnessed a test in the USA where ranges of ten miles had been observed from 10 000 feet. Although he thought that 20 miles from 20 000 feet could be achieved given another three to six months development work, Dee who had returned from his sick leave, emphasised that we were still in the research stage and that 'the present programme of 100 H_2S Mark VI equipments should be regarded as an expression of faith in the eventual importance of K-band.[11]

Mark VI was defined as an H_2S system using essentially the same computing and stabilising arrangements as those in Mark IIC and IIIA. That is, the main advance over those S and X-band systems was to be in the improved definition of the PPI picture. The introduction of more sophisticated navigation and bombing systems into the K-band H_2S, such as those devised for Mark IV H_2S, was in the process of formulation in TRE during the summer of 1944 and this advanced H_2S retained the code-name Lion Tamer at that stage.

The Mark VI K-band H_2S which first became airborne in Lancaster ND 823 from Defford on 25 June (1944) had one major new feature apart from the reduction in wavelength. This was the PPI Indicator (Type 216) which used an electromagnetically deflected CRT. The previous PPI indicators so far used in H_2S had electrostatic deflection with 4 kilovolt anode/cathode potential. Although the spot size on the tube did not interfere with the definition of the S and X-band H_2S systems, this was not the case for the higher definition given by K-band. The Type 216 Indicator used on Mark VI H_2S employed a high beam current electromagnetically deflected CRT with 7 kilovolt anode/cathode potential. Although the spot size no longer impaired the fundamental resolution of the H_2S a problem arose with the method of scan correction. In the electrostatic CRTs, voltage waveforms in quadrature were applied to the deflecting plates of the CRT to produce the rotating time base. In the electromagnetic CRT current waveforms had to be applied to the plates. There was no difficulty with a linear scan but a new method had to be devised for the scan corrected PPI. [12]

The high-definition of the K-band equipment also led to the introduction of a new feature on the PPI—the sector display. In the normal display the rotating time base has the centre of the CRT as its pivot. In Mark VI H_2S (on the 8 and 20 mile ranges) it was possible for the operator to switch this pivot to one of eight compass points around the edge of the tube and at the same time the amplitude would be doubled so that the full tube diameter was utilised.

The American RF systems used in these Lancaster flight trials comprised a magnetron delivering a peak power of 35 to 45 kilowatts. The pulse width was 0.25 μs and recurrence frequency 2000 per second. The magnetron trans-mitter was in a unit sealed at ground level pressure which also contained the local oscillator, crystal mixer, TR unit and intermediate frequency stages. The flight trials of this Mark VI K-band H_2S in the Lancaster gave ranges of about 19 miles from an altitude of 15 000 feet.

(iv) The 6 foot Whirligig, 3 centimetre H₂S Mark IIIC, IIIE and IIIF

At the second of the policy meetings on 16 June (1944) it was agreed that we would demonstrate the Lancaster installations of H_2S Mark IV and of IIIC (i.e. 3 centimetre 6 foot scanner—otherwise known as Whirligig) to Bomber Command at Defford between 28 June and 1 July. After endless difficulties with A V Roe in our attempts to get them to fit a cupola under the Lancaster for the 6 foot scanner, we designed one in TRE and it was on 4 July (1944) before Lancaster JB 558 first became airborne with the X-band 6 foot Whirligig system (H₂S Mark IIIC). The scanner of 6 foot aperture was fitted underneath the fuselage of the Lancaster and is shown with the 'camel blister' cupola removed in figure 27.1. The shape of this scanner departed significantly from the truncated paraboloid or barrel type scanner previously used. N Z Alcock (a member of my group from New Zealand) and Ramsay concentrated on the evolution of this 6 foot Whirligig scanner, but their description of it was not published until after the end of the war.[13] The reflector was cylindrical, 70 inches × 14 inches and the primary feed was an edge-slot waveguide array. The azimuth beam-width was 1.5 degrees. The elevation polar diagram corresponded reasonably well to the ideal cosecant[2] pattern from a depression of about 10 degrees to 75 degrees and there was a remotely operated two-position switch to change the maximum of the elevation beam from 3 degrees for navigation to 10 degrees for the bombing mode.

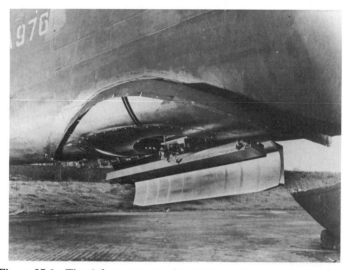

Figure 27.1 The 6 foot aperture slotted waveguide feed scanner used on H₂S Mark IIIC (Whirligig). The scanner was roll stabilised over ± 30 degrees.

On 6 July we gave Bennett an impressive demonstration of this 3 centimetre H₂S using the experimental 6 foot scanner. By that time we had made the problem of decision by Bomber Command even more difficult and complicated by offering additional versions, as follows:

(a) IIIC—the straightforward 6 foot scanner otherwise as Mark IIIA (that is roll stabilised, scan corrected indicator (Type 184) and track line).

(b) IIIE — improved IIIA with 3 foot scanner, using magnetic tube indicator (Type 216), adjustable tilt on scanner as in Mark IV (see p 218), pulse width reduced to $1/2\,\mu s$ with pulse recurrence frequency 1000 per second.

(c) IIIF — as IIIE but with 6 foot scanner.

In other words we were now offering in IIIE several of the major advantages of Mark IV by *ad hoc* changes to IIIA and this, together with twice the definition, in the IIIF alternative. At this stage, fortunately, having produced a summary of the S and X-band equipments already in service or being introduced (see Table 1 Chapter 24), I extended it to include the confusing miscellany of the new versions of H₂S. As one policy meeting succeeded another so we produced *ad hoc* changes in collaboration with Saward and his team which immediately generated another policy meeting. Table 2 bears the date 30 June 1944.

References

1 Smith J B 1985 *The new H₂Ss* in *Proc. IEE* **132** pt A p 404; originally published in TRE Jr. No 3 pp 78–97

2 Smith J B 1985 *The new H₂Ss* in *Proc. IEE* **132** pt A pp 404–406

3 Carter C J 1946 *H₂S: an airborne radar navigation and bombing aid* in *Proc. IEE* **93** pt IIIA 449

4 Smith J B 1985 *The new H₂Ss* in *Proc. IEE* **132** pt A p 406

5 Carter C J 1946 *H₂S: an airborne radar navigation and bombing aid* in *Proc. IEE* **93** pt IIIA p 464

6 Lovell TRE record p 33

7 Bean B R and Dutton E J 1966 *Radio Meteorology* National Bureau of Standards Monograph 92 Chapter 7

8 Most Secret document CS 15536 *Meeting to consider the policy to be adopted for future marks of H₂S* was held in Room 71/II Air Ministry, Whitehall on Tuesday 22 February 1944.

9 Lovell TRE record p 34

10 The Minutes of *A second meeting to consider the policy to be adopted for the future marks of H₂S* held in Room 71/II Air Ministry on Friday 16 June at 11.30 hours. Document CS 15536 is in Lovell TRE record at p 38

11 Corrigenda to paper CS 15536 in Lovell TRE record p 38

12 The details of the problem and the solution have been outlined by Smith J B 1985 *The new H₂Ss* in *Proc. IEE* **132** pt A p 409

13 Alcock N Z and Ramsay D W C *High level H₂S Scanner—Type 97* TRE report T2024 dated 3 July 1946

Mark	Brief description	Approx. present state of units	Approx. present A/c position	Policy
VI	K band using Sylvania RF Head. Same computing and stabilisation arrangements as in Mark IIC and IIIA. Uses magnetic tube Type 216 Indicator	200 Sylvania + 300 A.P.S.1 heads on order from U.S. 400 electrical units being made at RPU. Scanners (Type 82) by Nash & Thompson.	Experimental installation in Lancaster just beginning flight trials at Defford.	Installation on crash programme basis for PFF if system is successful during coming autumn and winter. Original target was 100 by end of 1944.
IIIC	X band 6ft array, otherwise same as IIIA	Nash & Thompson have development contract for Scanner (Type 86)	Has flown in Lancaster at Defford. Now converted to IIIF (see below)	
IIIE	X band. Improvements to IIIA without involving major changes. e.g. Magnetic tube Indicator Type 216 ½ microsec. pulse 1000 p.r.f. Improved scanner with perspex barrel and adjustable tilt. Automatically blacked out range marker.	Main changes to Mark IIIA units as follows:- Indicator 216 (will be in production for ASV VIB and H_2S (Mark VI) instead of Indicator 184. Modulator Type 174 (change of line in Modulator Type 64) Modifications to Scanner Type 71. (Power Unit 617 (will be in production for ASV VIB and (H_2S Mark VI) and Universal (Power Unit 567 (in production) instead of Power Unit 280. New Waveform Generator Type 46 instead of Type 34. Minor column 7 changes.	Experimental development complete at TRE. System of electrical units installed in Lancaster under IIIF.	
IIIF	Same as "IIIE" but with 6ft array as in IIIC.	Same as "IIIE" + Scanner Type 86	System installed in Lancaster at Defford	
IV	X band. Similar performance to IIIA but radically new computing and wind finding arrangements. Range stabilisation of picture. Automatic scan correction with altitude. Track and bombing markers automatically linked to Mark XIV Bomb sight.	Development at Gran.Co.complete. Type approval to be given shortly.	Gran.Co. prototype system now undergoing flight trials in Lancaster at Defford.	
Lion Tamer	K band (or X band) using Navigation and Bombing Computer	System in process of formulation at TRE.		

Table 2   New Developments, 30 June 1944.

Chapter 28

July 1944

At the second Renwick policy meeting of 16 June it had been decreed that a decision about the policy to be adopted with the various varieties of H_2S, described in Chapter 27, would be taken at a meeting which Renwick would convene in the week commencing 12 July 1944. Although we were a few days late with the flight demonstrations to Bomber Command this meeting did, in fact, occur at TRE on Sunday 9 July.

It is scarcely surprising that the demonstrations to Bennett on 6 July and to Saward, Donaldson and Thompson from HQBC on 8 July left impressions that depended largely on how well the particular equipment was working at the time of the flights. For example, on 6 July Bennett had four flights with various marks of H_2S, first with the Whirligig in Lancaster JB 558, secondly with Mark IV. Then he demanded to have a flight with H_2S Mark IV using the 6 foot scanner. Over the lunch break the Defford contingent of my group succeeded in converting JB 558 into a nearly complete Mark IV system using the 6 foot scanner but the apparatus failed after it became airborne with Bennett. Finally he wanted a demonstration of the K-band H_2S Mark VI.

By 8 July we had repaired the IIIF equipment and concentrated on demonstrating this and Mark IV. In effect this was a straightforward comparison of the best we could offer on 3 centimetres, using either the 3 foot or 6 foot scanners. Their reaction was surprising. The definition and contrast with Whirligig (IIIF) was the best they had ever seen but they complained that at short ranges the PPI picture of towns 'broke up' too much. They landed full of enthusiasm for Mark IV, which, considering it was one of the best demonstrations ever staged, was scarcely surprising. There had been time for J B Smith and the Mark IV party to produce a demonstration Lancaster with a judicious selection of CRT, and RF equipment and the display on the PPI of the convoy in Pembroke Dock with the Lancaster at 18 000 feet altitude (figure 28.1) was certainly revealing. Indeed, the engineered version of Mark IV produced results which were in a different category from those we were obtaining with the original H_2S Mark I when it became operational only 18 months earlier.

229

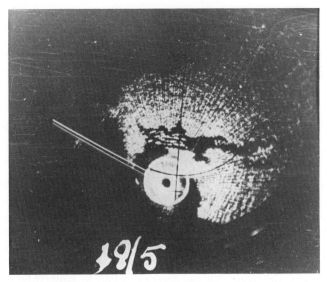

Figure 28.1 Photograph of the PPI display of H₂S Mark IV taken in a Lancaster flying at 18 000 feet over Pembroke Dock.

Meanwhile Dee had returned to TRE from his sick leave on 8 May and having replaced his stand-in, Denis Robinson, as Superintendent 3 there was an instant renewal of the acrimonious arguments about the various Marks of H₂S. Although he remained completely antagonistic to our Whirligig 6 foot X-band ideas he had softened to the extent of conceding that a 6 foot scanner installation on a Lancaster was practicable. This stimulated his idea that the really worthwhile advance would be a 6 foot scanner used on K-band with a full navigation and bombing computer system—Lion Tamer.

These topics were discussed endlessly and Rowe's attempt to achieve a degree of unanimity before the CCE (Renwick) meeting arranged for 9 July failed completely. In retrospect I realise that Dee, benefiting from his months of absence from the daily pressures, had the rational outlook of a scientist. The minute he sent to Rowe on 4 July summarised the situation precisely and is reproduced below in full.

> Over lunch to-day I felt compelled to send you a very short final summary of my dissatisfaction of our inability this morning to arrive at a unanimous opinion. To put it bluntly, Mark IIC and Mark IIIA H₂S are two new equipments which are scheduled to be introduced *in succession* to the whole of Bomber Command Main Force. PFF is to have Mark IIIA only as soon as possible. PFF are also scheduled to have a few squadrons of K-band (Mark VI H₂S), and much work on this in TRE and the US remains to be done. The whole of this programme is definitely laid on and scheduled to be complete by the end of the coming winter bombing season. I wonder, in fact, how much of this equipment will be in actual use by April next year?

A P Rowe with some of his senior TRE staff. The date of this photograph is probably 1943/44 when Rowe's office was the former 'classical sixth' room of Malvern College. From left to right J A Ratcliffe, R A Smith (standing), C Holt Smith, R Cockburn (standing with pipe), A P Rowe (with pipe) and his deputy W B Lewis. (Reproduced courtesy of Douglas Fisher.)

At the present moment not a single one of any of the above three equipments is available to Bomber Command. Surely this is a moment we should have seized in order to ensure that any following equipment is a first best (Liontamer) rather than a second best. Already this decision has been compromised in that the first best (Liontamer) has been forced into obscurity by the present discussion of two new equipments. It surely therefore is not asking too much that we should go for the second best (K-band with 6′ scanner), even if perhaps a month or two later, rather than a third best (Mark IIIF X-band 6′ scanner).

In my opinion it is completely wrong to excite interest, let alone tolerate a crash programme upon an X-band 6′ scanner system. This will certainly absorb effort within TRE and at contractors which could much more profitably be used to give Mark VI a 6′ scanner and make it into a system which then perhaps could reasonably be tolerated as an interim system pending the introduction of Liontamer. A bold man viewing the present state of the war in Russia, Normandy and Italy would probably say that neither Mark VI nor Mark IIIF should be the subject of crash programmes especially as two equipments are on the shelf pending production and introduction by Bomber Command. This morning's discussion was to me an appalling example of the manner in which we have become hack crash programme engineers, in that for three hours we argued the relative merits of two systems either of which is, at the best, only a small step towards what might now be regarded as the next reasonable objective.

However, none of us envisaged the enthusiasm of Saward and his colleagues for H₂S Mark IV after the demonstration on 8 July and the results of the CCE meeting on the 9th projecting our tasks into a future that, mercifully, was never to be, are summarised in my TRE record of that event.[1]

> Sunday July 9th 1944 ... H₂S policy meeting (at TRE) with CCE [Renwick] in the chair. Large gathering including DG of S [Director General of Signals], D of Radar [Director of Radar], DCD [Director Communications Development], Headquarters Bomber Command people and Bennett. The arguments were very extensive and in the main irrelevant to the conclusions of the meeting which were:
>
> 1. PFF
>   (a) PFF would get Mark IIIA as arranged this year.
>   (b) A crash programme on Mark IV as soon as possible for this winter, 6′ array to be added as soon as possible.
>   (c) Mark VI using a 6′ Scanner and if this did not have sufficient range for operational use it would be considered for low level precision work in Lancasters or Mosquitoes.
>   (d) Lion Tamer September/December 1945.
>
> 2. Main Force
>   (a) H₂S Mark IIC as arranged.
>   (b) To be followed by H₂S Mark IIIA as arranged with the proviso that improvements might be put in as and when possible during 1945.
>   (c) Main fitting of Lion Tamer 1946 January ...

Few people were satisfied by this, and in fact (except in the case of Mark IV) it made little difference to what happened. Many people, especially Bomber Command Headquarters were wrathful because a main production Mark IV hadn't been laid on. In fact many subsequent attempts were made to do so with little success. Flying Bombs so upset the country's economy that it was only with long delay that even the limited crash programme could be got under way.

The TRE view on this, was in the main, that it was too late and that one should go all out for Lion Tamer. To this end a separate team was quickly initiated with a programme to fly by October, but people were too tired, no firms could be persuaded to take contracts and this Mark VII and Mark VIII eventually fell into the limbo of 'peace time' projects.

References

1 Lovell TRE record p 43

Chapter 29

Naxos and H$_2$S

In the autumn of 1943 at about the same time as evidence accumulated that the U-boats were listening to the 10 centimetre ASV transmissions from Coastal Command aircraft, British intelligence sources concluded that German night fighters were homing on to the 10 centimetre H$_2$S transmissions from Bomber Command aircraft. For example, R V Jones[1] states that he had

> obtained overwhelming evidence that the Germans were plotting it [the H$_2$S transmissions] (they had started as early as November 1943), and were in fact equipping their fighters with a receiver code-named 'Naxos' to home on to the transmissions not only from H$_2$S but also on the kindred equipment fitted to our bombers to warn them of the approach of German night fighters. Since H$_2$S was switched on as soon as our bombers took off from their airfields in England, the Germans could get very early warning of a raid, and get their fighters airborne so as to be able to home on the transmissions when they came within range.

The development by the Germans of a device to detect the H$_2$S transmissions from the RAF bomber aircraft was scarcely surprising. Early in February 1943, a few nights after the first operational use of H$_2$S over enemy territory at the end of January 1943, a Stirling bomber fitted with H$_2$S was shot down over Holland. It crashed near Rotterdam during the night of February 2–3. The dismay of the Germans when they discovered that the bomber carried a system for generating high powers on centimetre wavelengths has been described by Pritchard[2] and by Price[3]. Both authors describe the confusion reigning in the German radar industry at that time and the difficulty their scientists found in discovering the function of H$_2$S which they had code named *Rotterdam*.

In this Stirling, and in several bombers subsequently destroyed over enemy territory, the navigator's PPI display had been destroyed and in spite of interrogation of members of the captured RAF crews it was 23 March (1943) before Colonel Schwenke, the Intelligence Officer in charge of the section dealing with captured enemy equipment, informed General Erhard

Milch the Director General of Air Equipment, that:

> The sets which have fallen into our hands have so far lacked their display unit
> ... but the interrogation of the prisoners has revealed that the device is
> certainly used to find targets, inasmuch as it scans the territory over which
> it flies...[4]

Three weeks after the crash of the Stirling, General Wolfgang Martini, the
Head of the German Air Force Signals Service had set up a *Rotterdam Commission* to develop any necessary countermeasures to H_2S. The captured H_2S
units had been handed over to the Telefunken Company but their investigation had been hindered by the destruction of the units captured from the
Stirling bomber during an RAF raid on Berlin on 1 March which severely
damaged the Telefunken works. Eventually the recovery of H_2S units from
other crashed bombers enabled Telefunken to assemble an H_2S system 'in
one of Berlin's huge concrete flak-towers, where it was safe from the heaviest
bombing attack. There its secrets were laid bare...',[5] and the development
of a simple detector for centimetre waves— *Naxos*—was commissioned. An
experimental model first became airborne on 11 September 1943 and proved
capable of detecting the H_2S transmissions at a range of about 10 miles.[6]

By 28 November (1943) the Germans had one fighter in operation
equipped with Naxos and,

> ... although it was currently proving impossible to distinguish between indi-
> vidual bombers with this device, the aircraft was still of use for shadowing
> the bomber stream and homing in the rest of the night-fighter force: it was
> capable of picking up H_2S aircraft in the bomber stream from over 60 miles
> away.[7]

By the spring months of 1944 several dozen German night fighters were
equipped with Naxos and a severe crisis had developed about the continued
use of H_2S by Bomber Command. The intelligence reports about the use of
Naxos by the German night fighters coincided with a period of heavy losses
to our bombers and in both the RAF and in TRE the issue was seized upon
by the opponents of H_2S to such an extent that at one point it seemed likely
that the use of H_2S by Bomber Command would be abandoned. There was
a division of opinion at all levels. On 8 May (1944) D of Tels (Director of
Telecommunications) and Renwick came to TRE to discuss the subject but
the matter was treated lightly and not very seriously. On 21 July, however,
Dalton-Morris the Chief Signals Officer of Bomber Command visited TRE
to complain about the heavy losses over the Ruhr and said that they
suspected the German night fighters were getting into the bomber stream
by using Naxos to home on to the H_2S transmission. Indeed, he said that
a captured German night fighter pilot had already claimed to have shot down
one of our bombers 'by using Naxos'.

After this there was a rapid polarisation of attitudes. Dalton-Morris was a well-known antagonist of H_2S and the relations between him and Saward were unfriendly. Dalton-Morris lined up with B G Dickens, Head of the Operational Research Section of Bomber Command and with amazing rapidity the situation became acute. Two days after Dalton-Morris' visit to TRE on 23 July Bomber Command was ordered not to switch on their H_2S during a raid on Kiel and most improbable stories immediately circulated amongst the Bomber Command crews about the danger of being shot down if they switched on their H_2S.

In TRE on 1 September R Cockburn, the Head of the Countermeasures Division, returned from several weeks liaison with Bomber Command and announced that H_2S would have to 'come out of the Command'. Shortly after, Cochrane the Commander of 5 Group forbade his crews to use H_2S entirely except for a few pathfinders. By that time wild accusations were being made on both sides and the situation deteriorated to such an extent that on 22 October (1944) Dee was persuaded by Rowe to spend three weeks investigating the situation at firsthand at Bomber Command and with the sources of the intelligence.†

Dee returned to TRE on 12 November from these investigations principally giving an enthusiastic account of the accuracies achieved by 5 and 8 Groups when using H_2S. On the subject of the danger of Naxos to Bomber Command aircraft he did not seem greatly perturbed, but thought that there might be some justification for keeping it switched off until about 5 degrees east by which time German ground-based radars would, in any case, have accurate plots of the bomber stream. Dee introduced a measure of sanity into the situation that had become dangerously emotional. He warned of possible future dangers of fighters homing on to H_2S aircraft and initiated development for 'storing' H_2S PPI information using skiatrons and other devices.

By that time the major panic about the dangers of using H_2S was subsiding. As related by Saward the claim made by Dalton-Morris during his visit to TRE on 21 July, that a captured German pilot claimed to have shot down a bomber 'by using Naxos', turned out to be a myth.

Squadron Leader W.H. Thompson of my staff was a fluent German linguist, and I arranged for him to interrogate the German pilot who had recently been shot down and made a prisoner of war and who was alleged to have made statements about the use of Naxos. Oddly, the Naxos equipment had not been installed in his aircraft, which was reasonably intact in its crashed condition. This interrogation took place at Trent Park, near Cockfosters in Hertfordshire, on 14 October. The prisoner of war explained to Thompson

† R V Jones[1] is rather vague about Dee's investigation, implying that he spent 'some weeks' with him and his intelligence group but he gives no dates. Dee certainly discussed the matter with R V Jones and his group but spent most of the three weeks at the Bomber Command groups.

what he knew of the system and how it was used, describing in some detail the method of presentation of the information it received. He also stated, most emphatically, that the device was designed to be used only to locate the bomber stream, the instrument being crude to the extreme, providing no measurement of range or accurate bearing of detected H_2S transmissions. This German went on to say that for attack the fighter relied on instructions from his Ground Night Fighter Control and the use of his radar interception equipment known as SN2, which was comparable to Britain's AI.[8]

Immediately after this interrogation Saward compiled a report at the request of Saundby on the operational use of H_2S in which he analysed the bomber losses on each target when H_2S was used. Contrary to the reports that had led to the anxiety about the use of H_2S, Saward found that during the period of most widespread use of H_2S the bomber losses had actually decreased from four per cent to under two per cent and that the introduction of the Fishpond attachment to H_2S was associated with a sharp decline in these losses. His report includes the comment that all the available information about Naxos had emanated from prisoners of war and no actual equipment had been captured.

> ... it will be seen that Naxos may be useful for homing into the stream, but has serious limitations as a 'homer' on to individual aircraft. It may therefore provide early warning of the approach of the Bomber Force, but since the plotting of the Force over enemy territory is already accomplished by other means, i.e. sound location, ground Radar plotting etc., crews need not be unduly intimidated by the German's use of Naxos ... The chief value of Naxos to the Germans may be as a propaganda weapon in an endeavour to stop, or at least limit, our use of H_2S...[9]

The antagonism and anxiety about the use of H_2S evaporated quickly shortly after Saward's report was circulated. A major influence in the renewal of confidence resulted from the capture of a complete German night fighter equipped with a device called 'Flensberg'.† This was designed to home on to the tail warning device Monica carried by the RAF bombers (see Chapter 25). Monica had been replaced by Fishpond in H_2S-fitted bombers but at that time there were many bombers, particularly in 3 Group which were not fitted with H_2S and still used Monica. Flight trials of the captured Flensberg by Derek Jackson revealed that it was a highly effective device capable of homing from a range of 50 miles to within 1000 feet of a bomber using Monica. 'As a result the Commander-in-Chief ordered the complete removal of Monica'.[11]

The abandonment of Monica and the acceptance of Saward's report that H_2S transmissions did not add significantly to the danger of attack by

---

† According to R V Jones[10] the JU88 carrying Flensberg landed in error at Woodbridge in Suffolk on 13 July.

German night fighters, restored confidence in the use of H_2S as a navigational and bombing aid and in Fishpond as a tail warning device. In fact at the meeting held on 9 February 1945 to discuss H_2S policy (see Chapter 32) Naxos was dismissed as of no consequence to H_2S policy in the following words:

> H_2S is accepted as a difficult system to jam, but by virtue of its continuous transmissions from an aircraft it is possible that it can provide a source of homing for the enemy. Whilst admitting this source of homing, past experience has shown that the enemy has so far been unable to utilise H_2S transmissions for effective homing, i.e. our losses for the past few months have been the lowest recorded in the history of the Bomber Force. [12]

A similar statement regarding the danger of Naxos is made by Webster and Frankland in their official history: [13]

> ... the conclusion reached in the course of 1944 by Bomber Command that an aircraft equipped with H_2S was, as an individual, in no greater danger than one without H_2S is correct.

References

1   Jones R V 1978 *Most Secret War* (London: Hamish Hamilton) p 392

2   Pritchard David 1989 *The Radar War: Germany's Pioneering Achievements 1904–45* (Patrick Stephens) p 88 *et seq*

3   Price Alfred 1977 *Instruments of Darkness, The history of electronic warfare* (Macdonald & Janes) p 134 *et seq*

4   See reference 3, p 136

5   See reference 3, pp 136–137

6   See reference 3, p 175

7   See reference 3, p 185

8   Saward Dudley 1984 *Bernard Lovell* (London: Robert Hale) p 113

9   Saward Dudley 1984 *Bernard Lovell* (London: Robert Hale) p 114

10 Jones R V 1978 *Most Secret War* (London: Hamish Hamilton) p 393

11 Jones R V 1978 *Most Secret War* (London: Hamish Hamilton) p 466

12 Report (BC/S 26829/Radar) of a meeting to discuss H_2S policy held at Bomber Command on 9 February 1945 [copy in Lovell, TRE record p 51]

13 Webster Sir Charles & Frankland Noble 1961 *The Strategic Air Offensive against Germany 1939–1945* vol IV (London: HMSO) footnote 2 p 13

Chapter 30

D-Day,
H₂S and the Army

Our long months of preparation and planning for the greatest amphibious operation in history ended on D-Day, June 6, 1944. During the preceding night the great armadas of convoys and their escorts sailed, unknown to the enemy, along the swept channels from the Isle of Wight to the Normandy coast. Heavy bombers of the Royal Air Force attacked enemy coast-defence guns in their concrete emplacements, dropping 5,200 tons of bombs ...[1]

Indeed, the disputes about H_2S related in the previous chapters were framed against an obvious change of priorities and interests. The division in TRE, led by Cockburn, responsible for countermeasures was the centre of interest and priority. Everywhere the concentration of interest concerned the moment when the Allies would invade Europe.

Sunday, May 28 (1944) The invasion of the west has still not come, but clearly one of the great moments of history is at hand. The roads are full of endless streams of tanks and lorries and the air of planes which stream out day and night over France and Germany.

As far as Bomber Command was concerned it was the short-range precision radars—Oboe and GH that held the limelight.

June 8, (about D-day) ... The air effort was extraordinary—11,000 aircraft being involved. Bomber Command alone despatched 1300 heavies and with the help of Oboe obliterated 33 out of 41 priority targets such as gun emplacements on the coast ...

H_2S had little part to play in these short range precision bombing sorties but, some days before D-Day, one of our H_2S equipped Halifax bombers at Defford became involved in an urgent operational test when it was feared that one of the complex deception manoeuvres designed in Cockburn's division might be compromised. This was a remarkable and complex electronic spoof designed to mislead the Germans into the belief that a major invasion

238

was occurring across the Straits of Dover. The spoof fleet involved only eight aircraft and four RAF air–sea rescue launches. The aircraft starting from Newhaven flew a transverse and slowly advancing pattern requiring precise navigation. With the electronic equipment in them and the sea launches and with an extensive use of jammers and the dropping of Window (aluminium foil reflectors) the German airborne radars reported that a massive invasion fleet was moving towards Cap d'Antifer and Boulogne.[2]

Of this spoof Churchill[3] wrote:

> There had been much argument [amongst the German High Command] about which front the Allies would attack. Rundstedt had consistently believed that our main blow would be launched across the Straits of Dover, as that was the shortest sea route and gave the best access to the heart of Germany. Rommel for long agreed with him. Hitler and his staff however appear to have had reports indicating that Normandy would be the principal battleground. Even after we had landed uncertainties continued. Hitler lost a whole critical day in making up his mind to release the two nearest Panzer divisions to reinforce the front ... It was not until the third week in July, six weeks after D-Day, that reserves from the Fifteenth Army were sent south from the Pas de Calais to join the battle. Our deception measures both before and after D-Day had aimed at creating this confused thinking. Their success was admirable and had far-reaching results on the battle.

This whole electronic deception operation could have been nullified if a single unjammed German radar had been operating. A report of a German radar working on a wavelength which would not have been subject to jamming was received some days before the invasion date. Martin Ryle, who was in charge of this countermeasures operation reminded me 38 years later of the emergency involvement of one of our H₂S aircraft in this episode.[4]

> 10 days before the original date for D-Day a report came through of a 'short-wave' (i.e. < 50 cm) German radar, which would have disclosed the rather complex deception operation. That was on the Thursday p.m.; on the Friday I wrote a memo giving 3 methods which couldn't be done in time, and one which might (identification of the radars and then low-level fighter attack). This would need an H₂S-fitted aircraft (none of the Group 5 [Ryle's Group] ones had H₂S) and help in the design and construction of a new feed–horn and waveguide filter. Approval came within a few hours and one of your people did the design and your workshop built the r.f. bits during Friday night, while we modified a broad-band receiver which existed, and studied the H₂S cabling to see how much could be adapted.
>
> The bits were all ready on Saturday morning and were installed and tested that day. I did a test-flight on the Sunday (May 27) (it was an important cricket event for the Defford RAF team—and a rather disgruntled Flt/Lt turned up in white trousers, blazer and spikey boots; he sent one of the ground-crew off to get a parachute and we took off!). Fortunately it all work-ed—giving a nice north-oriented display, and after tidying up cabling etc.,

was ready for an operational flight up-channel on the Monday night. Violent thunderstorms cancelled all flying, but it flew—with 2 navigators—on the Tuesday night—fortunately finding nothing to starboard. It was your Halifax HX166. I think it may have been a record for opl. requirement to opl. flight, —and I'm not sure how many other opl. flights were made from Defford. Anyway the Flt/Lt wasn't disgruntled any more!

H₂S and the Army

(i) H₂D

Although, apart from the incident related above, H_2S had no specialised part to play in the D-Day operations, other possible uses by which the army could be assisted had occupied our attention for many months. The existing means by which the movement of tanks and road convoys could be detected at night were somewhat crude—that is illumination by flares with visual identification or night photography. The possibility that H_2S could detect concentrations of tanks had often been discussed and in 1943 Killip made a number of flights using both 10 centimetre and 3 centimetre H_2S in collaboration with the army. However, he failed completely to find any evidence of radar responses from a concentration of tanks. The approach of D-Day gave a new urgency to the problem and we had hopes that our proposed high definition 3 centimetre and K-band systems would be more successful. On the basis of these hopes DCD gave us a general directive to do anything that was necessary.

This directive was first exercised in a most unexpected manner. On 30 April 1944 Robinson, who was then deputising for Dee, reported with enthusiasm that W B Lewis had just witnessed a demonstration by Kempton at ADRDE† of pulsed doppler detection of moving vehicles. The following day I visited Kempton with Thompson and by the end of that week we had installed a modified H_2S system in a trailer for operation in the pulsed doppler mode. Great Malvern was an excellent area for tests of such a system and on 4 May we towed the trailer to a position on Jubilee Drive where we could look down on trains moving along the line between Colwall and Great Malvern. The system detected the moving trains and so the next day we organised a more spectacular demonstration. On Sunday 6 May we towed the trailer nearly to the top of the Worcester Beacon and organised a convoy of vehicles to drive in and out of Malvern past the buildings which then served as the American Hospital and of which we had an uninterrupted view from our position on the hills. The system worked—with the pulsed doppler H_2S system the road convoys could be detected without difficulty.

---

† The Air Defence Research and Development Establishment then housed in Great Malvern a few miles from TRE.

Events then moved with great speed. The next day (7 May) we demonstrated the system to Rowe and by 8 May the H_2S Halifax HX166 was modified to a pulsed doppler system. Killip carried out the airborne tests and, as expected, found that the scanner had to be aligned fairly accurately along the line of flight otherwise the relative movement of the aircraft and the ground obscured the responses from moving vehicles. Nevertheless, after a test flight by Derek Jackson, he and Killip flew to London to meet CCE (Renwick), Leigh-Mallory† and Coningham.‡

> May 28 (1944) (reference these events)
> Work has been grinding along with one colossal flap — application of Doppler to detect movement on roads from H_2S planes. Demonstrated to an enormous number of people and CCE had a meeting ... the thing rather died, however, because nobody can think how to navigate along roads...

Even so the urgency of the problem was such that we were instructed to equip three Wellingtons, which we did by modifying three aircraft already fitted with ASV Mark III. The scanner mechanism was modified so that it was possible to 'inch' the scanner to any desired azimuth so that it could be set in the direction of the ground track of the aircraft to within a degree. These went to FIU (Fighter Interception Unit) for tactical trials. Then, from my TRE record[5]:

> This by no means ended the excitement, however, for on *Saturday May 13*, W/Cdr Eveley the CSO of 2 Group flew, on *Monday May 14* Pretty and the CO of 69 Squadron (Lousada) and on *Tuesday May 15* A.V.M. Embry the CSO of 2 Group. It is probably true to say that nearly all these people were impressed, but very much doubted if average operators could be trained to use the effect and also their ability to fly accurately enough along a road to make it useful.

All this occurred within the few weeks prior to D-Day and then the interest of the RAF, already overloaded with gadgets, diminished. We coded the system H_2D and with the 3 Wellingtons at FIU our work on H_2D steadily receded into the category of an 'experimental project'. It seemed unlikely that the H_2D system would be useful operationally because, with any simple form of presentation at least, it did not seem possible to discriminate the target from other spurious responses. H_2D was the first (and only) real 'non-starter' of H_2S and its various applications, but both CW and pulsed doppler systems became an important part of the navigational/bombing equipment of post-war H_2S-equipped bombers (see Chapter 32).

---

† Air Chief Marshal Sir Trafford Leigh-Mallory, Air Commander in Chief of the Allied Expeditionary Air Force.
‡ Air Marshal Sir A Coningham, Commanding Officer of the 2nd Tactical Air Force.

(ii) Six-foot Scanners on K-band

Although the pulsed doppler H_2D system made no progress towards an operational system the further development of conventional forms of H_2S soon showed great promise. After D-Day there was widespread public optimism that the war would be over by September. The Germans had been cleared from the beaches and Bayeux, five miles inland, had been captured but the warnings of Churchill were salutory.

> Thursday June 8th [1944] ... Rommel is now said to be moving up his reserves and Eisenhower and Churchill have reminded the country again today against over optimism and of the tasks ahead ...

Indeed, the belief that the war would end in the autumn soon vanished and once more we directed our efforts to find means of assisting the army. Although efforts to detect tanks by the H_2S Mark II and III systems had failed we had hopes that the far greater definition provided by using a 6 foot scanner on K-band would be more successful. By the first week in December (1944) we had this system ready for trials in Lancaster JB 558 and when flying at low altitudes (1000–2000 feet) the results were 'immediately staggering'. On the PPI in those first flights railways, runways, flooded fields, narrow streams and bridges could be seen and a new era was opened for H_2S. On 6 December (1944) Renwick arranged to fly photographs of the PPI, such as those in figure 30.1 to Montgomery.

The Mark VI K-band using a 3 foot scanner was about to be fitted in the

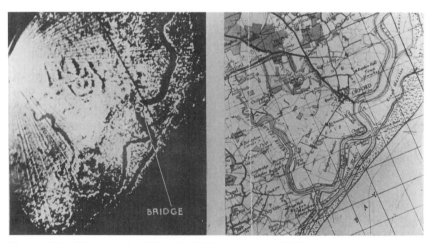

Figure 30.1 The PPI of a K-band (1.25 centimetre) H_2S using a 6 foot scanner photographed in a Lancaster flying over Suffolk in December 1944 at an altitude of 1000–2000 feet. The river Butley and a temporary bridge over the river Alde are well defined.

Bomber Command aircraft (see Chapter 27) but it was not giving really useful results at high altitude and we urged that the programme should be switched to use the equipment with 6 foot K-band scanners for Army collaboration.

> Saturday December 16th 1944 ... 10.00 hours policy meeting on K-band without an agenda, Saward representing Bomber Command. SAT† also there and an Air Commodore Cross (D of Ops.Tac) called in half way through. After a very long discussion and after travelling over much thin ice, the meeting agreed to the programme which we wanted, namely to proceed with the installation of the Mosquito for the low level work and to change the 3ft high level Lancasters from Bomber Command to 6ft low level Lancasters ...[6]

In order to compensate Bennett and his Pathfinder Force for the loss of the K-band Mark VI we proposed that he should be given 100 unstabilised 6 foot X-band systems with the new magnetic indicator (Type 216) $\frac{1}{2}\mu s$ pulse and high PRF (that is Mark IIIF). We had presented ourselves with a major problem of reorganisation of the manufacture of the scanners and aircraft fitting programme, nevertheless it seemed that we were on to a major advance in the adaptation of H₂S for Army use. On 5 January (1945) Pretty, at the Air Ministry, arranged a meeting with a number of Army Generals at which we made arrangements for flight trials of the equipment against tanks, pontoon bridges, and road convoys.

> ... and on Wednesday January 10th (1945) when England was covered in snow, 14 RAF officers arrived in Lovell's office hoping to take part in the trials and Embry sent a message that he wanted a demonstration at 11.30 on the morrow. However, Killip had sidetracked them, stranded at Newmarket in snow, and done the trials without them ... the tanks which were arranged in groups of 4, 20 yards spacing with 100 yards between the groups, were seen as isolated responses in the open and could be seen lining up and proceeding into the wood. They disappeared, however, in the wood.‡[6]

The clear detection of tanks by this H₂S equipment delighted us but surprised many others to such an extent that they maintained that the clarity by which the tanks could be seen on the PPI was because the ground was covered with snow. It was five weeks later before the snow cleared sufficiently for us to demonstrate the error of this belief by repeating the exercise with the tanks on bare ground. Although we referred to this as a 6 foot scanner the aperture was 5 feet 6 inches giving a fan beam with an azimuth

---

† Watson-Watt was jointly Scientific Adviser on Telecommunications (SAT) in the Air Ministry and VCCE (Vice Controller of Communications Equipment) in MAP.
‡ In addition to the Lancasters fitted with the 6 foot K-band H₂S, tests were also made with a Lancaster fitted with a 6 foot X-band H₂S but this equipment saw the tanks only intermittently.

Figure 30.2 A mosquito fitted with a 28 inch K-band scanner.

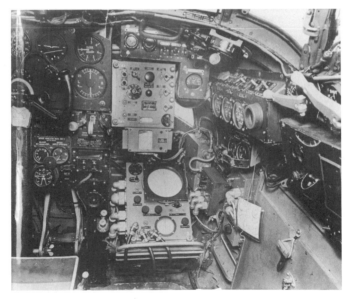

Figure 30.3 The cockpit of a H_2S-equipped Mosquito showing the H_2S PPI and control unit.

width of only 5/8 degrees. A detailed description of this scanner (known as Type 99) was issued as a secret TRE report[7] on 17 July 1945.

After these demonstrations it was agreed to fit six Lancasters with the 6 foot K-band H₂S and six Mosquitoes with a 28 inch K-band scanner (Type 96).[7] A Mosquito† fitted with this scanner is shown in figure 30.2 and the arrangement of the cockpit with the H₂S PPI tube is shown in figure 30.3. It had already been agreed that the Mosquitoes should be used by 5 Group for this low level marking but there was trouble with the organisation for using the Lancasters. Renwick eventually persuaded Harris to create a special wing in 3 Group under Flt/Lt Day (with whom, as a member of Saward's team, we had been closely associated). It was intended that this wing, still under the control of Bomber Command, would carry out tactical operations for the Army. All this was set in motion but before Day had time to train his crews the war in Europe ended and for security reasons D of Radar (Director of Radar in the Air Ministry) laid an embargo on the use of K-band equipment. However, before VE day 5 Group had already carried out three successful marking raids with their K-band H₂S-equipped Mosquitoes using the 3 foot scanner.

References

1 Churchill Winston S 1954 *The Second World War* vol VI *Triumph and Tragedy* (London: Cassell) p 3

2 For the details of this spoof invasion operation, see Pringle J W S 1985 *Proc. IEE A* **132** 340

3 See reference 1, p 10

4 Sir Martin Ryle to Lovell—letter dated 5 November 1982, deposited with the Lovell papers in the John Rylands University Library of Manchester. See also Ryle M 1985 *Proc. IEE A* **132** 438

5 Lovell TRE record p 44

6 Lovell TRE record p 45

7 Garfitt R G *Low Level H₂S K-band scanners Types 96 and 99* TRE report T1896 17 July 1945

† This particular Mosquito ML 910 broke its undercarriage on landing after its second flight and had to be written off.

Chapter 31

The U-Boat Schnorkel

In Chapter 19 I wrote that by mid-1944 when we had succeeded in developing an experimental version of the high-powered 10 centimetre ASV with lock–follow (Mark VIA), the battle against the U-boats in the Bay of Biscay was 'already history'.

Although the U-boat campaign never recovered from these setbacks of 1943 and the Allied liberation of the French West coast ports in 1944 there was, in fact, a worrying development in 1944. The Germans had fitted a breather tube known as *Schnorkel*, which enabled the U-boats to charge their batteries when submerged. In 1940 they captured two Dutch submarines fitted with a Schnorkel but attached no importance to its development until the heavy losses of 1943 when they were forced to search for means of survival for their submarines. Successful trials were made in July 1943 and by mid-1944 30 Schnorkel U-boats were operational. The projecting breather tube presented a very poor target for ASV; in fact, the ranges on Schnorkel in the presence of the radar clutter produced by a rough sea were operationally negligible.

The anxiety about the Schnorkel-fitted U-boats increased when the surge of the Allied armies across Europe was halted and it became known that the Germans were developing a new type of very fast submarine. Known as the Walther boat, this design, in addition to the normal means of propulsion, had turbines driven by gases produced from the combustion of diesel fuel and hydrogen peroxide—and was capable of very high under-water speeds for short periods. [1]

Of the anxieties caused by these two developments in 1944 Roskill[2] wrote

> By reducing the effectiveness of our airborne radar, the Schnorkel struck a heavy blow at Allied anti-U-boat tactics. Taken with the construction of high-speed submarines, on which we knew the Germans to be actively engaged, it was plain that unless victory could be gained before the new developments had come into general use, we might find ourselves struggling against an enemy who was once again possessed of the inestimable benefit of the initiative.

On 10 October (1944) E J Williams (a member of Blackett's operational

246

research staff at the Admiralty) came to TRE and asked what we were intending to do about increasing the range of ASV detection of the Schnorkel and on 22 October Rowe held an internal inquest in TRE to find out why we had failed to 'ring a bell' about the Schnorkel problem. As far as ASV was concerned we had no immediate prospect of improving on the powerful 10 centimetre ASVs which we had by that time developed. These—ASV Mark VI with the 200 kilowatt transmitter and attenuator, and the lock–follow and blind bombing ASV Mark VIA and VIB—seemed the only possible avenue of progress within the next six months at least. We tried some flights against a simulated Schnorkel with the experimental K-band equipment (see Chapter 27) but without much success.

At the end of October Dee was summoned to a meeting with the Prime Minister and on 22 November Renwick held an emergency meeting to hear what TRE intended to do about improving ASV ranges on the Schnorkel. Beyond ASV Mark VIA and VIB the unanimous opinion in TRE by that time was that the solution, if any, would come from special developments of a 3 centimetre (X-band) ASV. Since the early stages of the introduction of H_2S in the Pathfinder Force, O'Kane had been largely occupied in operational liaison with the RAF Commands and subsequently with the British Branch of the (MIT) Radiation Laboratory which had been established within the confines of TRE. When the ASV/Schnorkel problem emerged the need for this liaison was diminishing and, since any possible ASV solution was beyond the scope of further modification to the H_2S systems, this new problem was handed over to a new group under him.

The problem of retrieving the useful surfaced U-boat ASV ranges on the Schnorkel required entirely new thinking and, as R V Jones wrote:[3]

> Both in the air, with the ME 262 jet fighter, and at sea with the 'Schnorkel' U-boats and the fast Walther U-boats propelled by hydrogen peroxide, and with new homing torpedoes, the Germans would have had an impressive armoury if they had been able to sustain the war for another year or two.

The severity of the danger, had the Germans been able to sustain the war, is underlined by Churchill's description[4] of the continuing menace of the U-boats after the decisive attacks in 1943 against those operating from the Bay ports.

> The whole Anglo-American campaign in Europe depended upon the movement of convoys across the Atlantic, and we may here carry the story of the U-boats to its conclusion. In spite of appalling losses to themselves they continued to attack, but with diminishing success, and the flow of shipping was unchecked. Even after the autumn of 1944, when they were forced to abandon their bases in the Bay of Biscay, they did not despair. The Schnorkel-fitted boats now in service, breathing through a tube while charging their batteries submerged, were but an introduction to the new

pattern of U-boat warfare which Doenitz had planned. He was counting on the advent of the new type of boat, of which very many were now being built. The first of these were already under trial. Real success for Germany depended on their early arrival on service in large numbers. Their high submerged speed threatened us with new problems, and would indeed, as Doenitz predicted, have revolutionised U-boat warfare. His plans failed mainly because the special materials needed to construct these vessels became very scarce and their design had constantly to be changed. But ordinary U-boats were still being made piecemeal all over Germany and assembled in bomb-proof shelters at the ports, and in spite of the intense and continuing efforts of Allied bombers the Germans built more submarines in November 1944 than in any other month of the war. By stupendous efforts and in spite of all losses about sixty or seventy U-boats remained in action until almost the end. Their achievements were not large, but they carried the undying hope of stalemate at sea. The new revolutionary submarines never played their part in the Second World War. It had been planned to complete 350 of them during 1945, but only a few came into service before the capitulation.

Early in 1945 O'Kane equipped trailers with both X and K-band systems for his Schnorkel tests. By using a sector scan, high pass filters and by decreasing the pulse width to 0.5 microseconds (μs) and increasing the repetition rate he made tests on a Schnorkel from the trailers situated on the Great Orme. The European war ended before airborne tests could be made against simulated Schnorkels with any of these modifications. Simultaneously the problem of increasing the Schnorkel detection range was investigated by the Americans. According to Guerlac[5] flight tests revealed that X-band was more successful than K-band. At the request of the US Navy, the MIT Radiation Laboratory initiated *Project Hawkeye* in which various modifications were made to the US H_2X radars including the use of an 8 foot $\times$ 3 foot scanner. Although such modifications coupled with increased experience led to some success against U-boats equipped with Schnorkel, there is no doubt that, had the war continued, the problem of U-boat detection and attack would, once more, have become heavily dependent on surface vessels, for as Guerlac[6] concludes:

In the anti-submarine war, aircraft were a new and highly effective weapon. Much of this effectiveness must be attributed to radar. One of the most important results of the widespread use of radar, both airborne and shipboard, was in driving the U-boats beneath the surface, blinding and partially immobilizing them. At the end of the war the Germans had largely reduced the threat of air power by the use of Schnorkel but this, like all other countermeasures to radar, reduced the mobility of the U-boat and lessened its destructive power. Without radar linked with air power the U-boat menace could never have been reduced sufficiently to permit the building up of forces for the offensive operations which finally resulted in the defeat of the Axis powers.

The anxiety about a resurgence of the German U-boat campaign continued until nearly the end of the war. Indeed, early in 1945 the First Sea Lord sent a gravely worded memorandum to the Chiefs of Staff Committee. He anticipated a renewed offensive on a substantial scale in February or March, with large numbers of the new type of U-boats loose on the Atlantic convoy routes as well as in our coastal waters. Shipping losses might, he considered, even surpass those suffered in the spring of 1943; and if that happened the land operations in Europe were bound to be adversely affected. About this memorandum Roskill[7] comments

> ... In retrospect it may seem surprising that such grave forebodings should have arisen so near to the end of the war; but the menace represented by the new U-boats and especially by the 1600 ton Type XXI was very real—if they got to sea in large numbers, and the Admiralty could hardly have been aware of the full extent to which our bombing raids and air minelaying had disrupted the German progress for the completion of the new boats and training their crews.

It is fortunate that although the Schnorkel U-boats became operational in mid-1944 the greater danger of the Schnorkel plus the revolutionary high-speed submarines never materialised. In fact the design of the Walther boat suffered from long delays and troubles. Although seven small boats (320 tons) of that class were completed and two large prototype boats (1600 tons) had been ordered for training and experimental purposes, no operational version (850 tons) was sent on active service before the end of the war. According to Roskill[8] the problems with the Walther boat led Hitler in July 1943 to order priority for the Type XXI U-boats with the streamlined hull of the Walther boat, with batteries of greater capacity to give a bigger cruising range and capable of short bursts of speed of up to 17 knots under water. The planned production rate of 33 per month by September 1944 was a serious potential threat to Allied shipping but the disruption of the German industry by Allied bombing was such that according to Roskill[7]:

> ... by the end of January 1945 the Germans realised that only 1 or 2 Type XXI, instead of the planned 40, would be ready for operations in February and larger numbers would not be available until April. Biggest factor in delaying production was the breaching of the banks of the Dortmund— Ems and Mittelland canals by Bomber Command Lancasters in Nov. 1944— U-boat sections being too big to go by road to the yards at Hamburg and Bremen. The output dropped from 14 in October to 5 in November 1944.

References

1   Roskill S W 1956 *The War at Sea 1939–1945* (London: HMSO) vol 2 p 207

2   Roskill S W 1961 *The War at Sea 1939–1945* (London: HMSO) vol 3 pt 2 p 68
3   Jones R V 1987 *Most Secret War* (London: Hamish Hamilton) p 464
4   Churchill Winston S 1954 *The Second World War* vol VI *Triumph and Tragedy*
    (London: Cassell) p 472
5   Guerlac Henry E 1987 *Radar in World War II* vol 8 *The History of Modern Physics
    1800–1950* (Tomash Publishers and the American Institute of Physics) p 727 *et
    seq*
6   See reference 5, p 729
7   Roskill S W 1961 *The War at Sea 1939–1945* (London: HMSO) vol 3 pt 2 p 290
8   Roskill S W 1960 *The War at Sea 1939–1945* (London: HMSO) vol 3 pt 1 p 17

Chapter 32

The Last Months of War and the Post-war Phase

In the previous chapters I have described our efforts to develop an H_2S after D-Day which could assist the army. However, throughout those autumn and winter months, the problems with the various marks of H_2S destined for fitting in Bomber Command continued and the political issues were exacerbated by our involvement in the conflict between Bennett and his Pathfinder Force (8 Group) and Cochrane with his 5 Group.

After the TRE démarche of April 1944 (see Chapter 26) two of Bennett's squadrons were handed over to 5 Group. This brought the AOC of 5 Group, Air Vice Marshal The Honourable Ralph Cochrane, into our orbit. According to my diary, he 'looked and behaved exactly like Bennett' and on 2 August and again on 28 September (1944) he swept into my office 'full of enthusiasm for his marking methods and pleading for first class equipment in six of his aircraft'. As described in Chapter 26 we had complained about the deterioration in the use of our latest versions of H_2S by Bennett, and Cochrane's idea of a small highly trained precision marking force using H_2S was precisely the concept for which we were arguing. At that stage in the war a large section of my group was permanently stationed at Defford and we had considerable resources of aircraft and equipment at our disposal. It was not difficult for us to meet Cochrane's request and we quickly converted several of his Lancasters to H_2S Mark IIIE (ie the best available form of roll stabilised X-band using 3 foot scanners) and we agreed to fit K-band in some of his low-level Mosquito marking force. As related in Chapter 30, in the event, these transpired to be the only K-band H_2S used operationally before the war ended.

Our private enterprise with Cochrane and 5 Group soon became known to Bomber Command and the Air Ministry and we were strongly condemned. That did not affect us but the fact that Cochrane now had some of the best available H_2S in 5 Group greatly increased the antagonism between him and Bennett. I wrote in the diary at that time:

The war between 5 Group and 8 Group seemed at times to be more bitter than against the common enemy; however, the Germans suffered for it, because in the thick of competition the results of each improved beyond recognition.

On 7 September Bennett came to TRE to complain to Rowe and Dee that we were not giving H₂S enough attention. Neither were impressed by his complaints nor by his abrupt departure after 40 minutes conversation because he had to 'fly back to London for lunch'. Even so, our complaints in April and the intense competition between the two Groups had the desired effect on the performance of the Pathfinder Force.

> Tuesday November 21st 1944 ... Went with Barton and Whitbread after-wards to one of the newer PFF Squadrons (582 Squadron at Little Staughton). Cribb, now a G/C was the Squadron Commander there. At 10 o'clock the Squadron was just returning from Aschaffenburg. This raid was laid on as a master bomber raid to be marked by Oboe Mosquitoes, illuminated by flares dropped blind from H₂S Mark III and assessed by the master bomber, who corrected the Oboe marking if necessary. There was 10/10ths cloud over the target however. The master bomber gave 'upside down' (which means that all planes were to bomb on instruments), and hence the raid effectively became a blind bombing H₂S raid. The crews from this Squadron appeared to be well satisfied with their performance of H₂S and had no complaints regarding fades.
>
> ... There were complaints about the target breaking up 'as usual' on the 10/10 scale. The crews were mostly annoyed, however, about the instruction to keep their H₂S Mark III switched off until 5° E.† I gathered that Bennett had received a direct order from the C-in-C a few days ago to obey the instructions of the Command in this matter, even to the extent of switching off X band equipment. The crews complained bitterly that it did not give them enough time to tune up... The crews said that the opposition on this raid was quite negligible. One man had a Fishpond contact and successfully took evasive action. Fishpond has apparently come considerably into favour since the decease of Monica†... This Aschaffenburg raid only represented a little of Bomber Command's activities on Tuesday night. PFF led at least 3 other 200 bomber raids; 5 Group carried out a quite independent show, and there were the usual Mosquitoes and other oddities over Germany... .[1]

Indeed, during those autumn months of 1944 it was evident that H₂S was being used consistently and effectively. Towns deep in Germany, and hith-erto largely immune, began to look like Bremen, Stettin and Königsberg.‡ On 6 December Whitbread from 8 Group agreed to speak to the senior members of my group about the results they were achieving with H₂S. He said that over the last five months 53 per cent of their markers and bombs

---

† See Chapter 29.
‡ This reference is to reconnaissance photographs of these towns in my TRE record in which the central areas appear completely devastated.

had gone within one mile of the target point and 90 per cent under two miles. There had been only one instance of a gross error (greater than three miles from the target).

A few days later (on 10 December 1944) Day, our previous Liaison Officer with Saward, and now a Squadron Leader of 139 Squadron, told us that he had just completed his second tour and had done 36 operations with H_2S Mosquitoes. 25 of the sorties had been with S-band H_2S and 11 with X-band all from 25 000 feet or above. He had achieved a probable error of only 0.8 nautical miles (n m) and in his last raid on Berlin using X-band H_2S he was only 0.5 n m from the aiming point. He claimed that his Mosquito Squadron dropped a greater tonnage of bombs in one month than an equivalent Lancaster Squadron and that his Squadron had been able to operate on 22 nights during November. As evidence of the precision of his squadron he said that in their last raid on Nuremberg 70 Mosquitoes had done more serious damage than a previous raid with 400–500 heavy bombers. Not unexpectedly he was desperately anxious to get more X-band H_2S in his Mosquitoes (at that time only three were equipped with X-band).

In 1944 a squadron of the Mediterranean Allied Air Force (MAAF) equipped with H_2S Mark II was operating from Italy. The report[2] on these operations stated that a number of raids were 'converted from potential failures to successes by use of the equipment. Illumination of the targets was excellent and the main force squadrons were duly impressed'.

In spite of this widespread evidence of the success of H_2S some opponents remained. Early in 1945 having failed to suppress H_2S on any technical or tactical counts these opponents commenced an agitation that the manpower situation in the country was too poor to handle it and that Loran† should be substituted! This led to the last of the major policy meetings that I attended.[5] In relation to the manpower problem and the possibility of using Loran the meeting discussed the complete introduction of the various marks of H_2S to Bomber Command, and the appropriate versions of H_2S to send to the Far East.

At the time of this meeting I was already ill—an illness soon diagnosed as complete exhaustion and before the end of February I was sent away from TRE to recuperate. I did not return until after VE day. Although I was not officially released from TRE until 19 July (1945) I took little further part in the H_2S story. Both my personal diary and my TRE record ended at the time of the February policy meeting and for the completion of this account of H_2S I have depended heavily on the official records. In particular, I have been provided with invaluable details by Air Vice-Marshal P M S Hedgeland who

---

† A long-range navigational aid based on the principles of Gee, initially working on a frequency of 2 megahertz. After the war Loran was further developed and widely used for civilian air and ship navigation. For the history and details of the development of Loran in the USA see Bowen[3] and Guerlac.[4]

Part of the H_2S group photographed on the roof of the Preston Laboratory of Malvern College, February 1945. Seated in the front row reading from the left—H G Hinckley, F C Thompson, the author, Flying Officer J Richards and Fl/Lt Killip. Standing immediately behind the author Fl/Lt D W C Ramsay with Fl/Lt P Hillman behind Ramsay's left shoulder.

was with my group as a junior RAF Officer on detachment from Bomber Command at various times from the autumn of 1942 and remained associated with H_2S in the post-war years: as Radar Trials Officer at BDU/Central Bomber Establishment from 1945 to 1948 and as the RAF Project Officer for H_2S Mark 9A from 1952 to 1960. Miss Janet Dudley, the Senior Librarian at RSRE (formerly TRE) has searched the archives for the contemporary TRE progress reports and W H Sleigh, formerly Aeronautical Engineer on the staff of RSRE, has provided information about the post-war installations in the bomber aircraft. The following account is largely based on the information from these three sources.

VE Day to VJ Day

After the end of the European War on 9 May the nature of the requirements for H_2S in the Pacific dominated the developments. Evidently the war in the Pacific would be fought over much greater distances than in Europe and air-

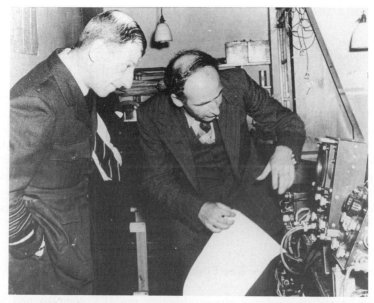

The author discussing a captured German copy of H_2S equipment with the Chief of the Air Staff, Marshal of the Royal Air Force Sir Charles Portal in the Preston Laboratory of Malvern College June 1945. Sir Robert Renwick is partially obscured. (Shortly after this photograph was taken Sir Charles Portal was raised to the peerage and in 1946 was created Viscount Portal of Hungerford, KG, GCB, OM. Sir Robert Renwick Bt, was made KBE in 1946 and was raised to the peerage in 1964.)

craft could no longer depend so heavily on ground-based aids. In the case of H_2S this implied the need for a system which would reveal coastlines at the greatest possible range.

The decision to equip the main bomber force with H_2S Mark IV was modified in favour of an improved version to incorporate a 5 foot 6 inch scanner, an indicator using electromagnetic deflection as the PPI and linked with the bombing computer and air position indicator.† However, it was not believed that production of this Mark IVA could commence until 1946 and interim measures were proposed for the VLR (Very Long Range) or Tiger Force destined for the Pacific. According to Hedgeland[7] the 10 centimetre Mark IIC system was to be converted to operate on 3 centimetres.

It was eventually decided that a simplified version of H_2S Mk IIIA would suffice, to be known as Mk IIIG, in which Mk IIC systems straight from the production line which was still running were converted to the Mk IIIA

† For the details of this linkage see Carter[6].

standard, less the stabilised scanner platform. It was also modified to give a 40–80 nm scan for long range detection of coastlines. Furthermore, every box was studied in detail and components liable to succumb to heat and humidity were replaced. These units were labelled 'MFP' meaning 'Modified for Pacific'. In the event 'Tiger Force' was never deployed to Okinawa as planned.

Another version known as H_2S Mark IIIH was also flight-tested in TRE during the summer of 1945. This was essentially the same as IIIG but used the improved electromagnetic indicator Type 216. At this stage there were 12 versions of H_2S either in operation or in various stages of modification and production. With the surrender of Japan in early August the immediate operational requirements for many of these versions of H_2S vanished. All marks of H_2S up to IIIG were declared obsolescent and work on them ceased in TRE.

The Post-war Phase

In the immediate post-war phase, although the H_2S complex had been greatly simplified by the decision that all Marks to IIIG were obsolescent, nevertheless six versions remained on the active and development list and it will be convenient to describe the progress with these in sequence.

H_2S Mark IIIG

The 3 centimetre H_2S Mark IIIG remained the standard version for the Lancaster bombers until it was superseded by H_2S Mark IVA (see below). Various improvements were incorporated such as the pressurisation of the modulator.

H_2S Mark IV/IVA

The production of the original Mark IV H_2S had been cut to 150 units when it was decided to equip the main force with an improved version H_2S Mark IVA using the greater definition provided by the '6 ft' (5 foot 6 inch) scanner. According to Hedgeland:[7]

> H_2S Mk IV did not go into production but the impending arrival of the Lancaster Mk VI, renamed the Lincoln, gave the opportunity to introduce an improved H_2S system. This was based on the Mk IV design with all its advantages derived from linkages with the Air Mileage Unit and Air Position Indicator to provide a ground stabilised display with a wind-finding capabi-

lity (always the biggest navigational problem) and the Mk XIV Bombsight. However, the improvements in range and definition provided in Mk IIIF were incorporated to give a 6ft roll-stabilised scanner and an indicator with magnetic deflection. This system was called H$_2$S Mk IVA.

The flight trials of the IVA system had been carried out from Defford by TRE throughout 1946 and the production line of the H$_2$S units commenced early in 1947. Hedgeland[7] recalls his association with Mark IV and IVA.

I moved from TRE to BDU at Feltwell in September 1945 and was shortly followed by Hillman. I recall that we did some trials on an H$_2$S Mk IV aircraft but in 1946 and 1947, when BDU had become the Development Wing of the Central Bomber Establishment at Marham, we conducted full Service Trials of H$_2$S Mk IVA on four Lincoln aircraft, including tropical trials based on Khartoum. The Lincoln with H$_2$S Mk IVA became the standard heavy bomber in Bomber Command until it was replaced by the V-Bombers (Valiant, Vulcan and Victor) in the late '50s.

H$_2$S Mark VI (K-band)

The K-band systems developed for low-level reconnaissance are described in Chapter 30. After VE day these were transferred to the BDU and trials continued until the end of 1945 in collaboration with the Army.

H$_2$S Mark VII (K-band)

Mark VII was the final version of the Lion Tamer project described in Chapter 27. The system was abandoned. First, the K-band radio frequency units were of American manufacture and with the cessation of the lease–lend arrangements this source of supply was no longer available. Second, although the K-band Mark VI system had operated satisfactorily at low level it was soon discovered that at higher altitudes under certain weather conditions there was sometimes complete loss of signal because of the absorption by water vapour (see footnote p 233).

H$_2$S Mark VIII

Although the Lion Tamer Mark VII K-band system was abandoned the concepts of integration with the aircraft's navigation and bombing computer (NBC) were carried forward to a 3 centimetre (X-band) version known as H$_2$S Mark VIII. In the autumn of 1945 it was decided to produce a dozen for trials, but subsequently the number was reduced to four. Fitted in a

Lincoln aircraft, TRE carried out trials of this system late in 1948 and gained valuable experience for the development of the ultimate H_2S (Mark IX) for the V-bombers.

H_2S Mark IX/IXA

The final version of H_2S working on 3 centimetres and the only H_2S to be designed as an integral part of the bomber. Attention was first given to this system in TRE in the autumn of 1945 in response to the Air Staff requirement for H_2S in the English Electric E3/45 jet propelled bomber later to be known as the B3/45—the Canberra. Whereas all the wartime H_2S systems had been added to existing designs of bomber with the scanners in an external cupola or nascelle, it was now to be designed as an integral part of the bomber with no external protuberances. From the beginning TRE envisaged this Mark IX H_2S as a further development of Mark VIII fully integrated with NBC to give automatic ground position and speed, wind, sector scan and offset bombing computation. Early in 1946 a formal recommendation was made that the system should be on X-band (3 centimetres) using a 200 kilowatt magnetron. In the spring of 1946 a development contract was accepted by the EMI/Gramophone Company and it was hoped that flight trials in a stripped Lincoln could begin in 1948 and in the E3/45 jet bomber in mid-1949.

However, this plan went awry because of the Air Staff decision in the early autumn of 1948 that this new H_2S should be designed for the B14/46 bombers (subsequently known as the V-bombers). More space was available for the H_2S in these aircraft and the consequent re-design of the H_2S (to Mark IXA) caused a revision of the programme. In the event flight trials of the original H_2S Mark IX units with the NBC took place during 1950 in a Lincoln aircraft.

The first engineered set of revised units (Mark IXA) destined for the V-bombers were flight tested from Defford in the early summer of 1951 and the trials continued either in a Hastings or Ashton aircraft† until the V-bombers were available. In 1956 this H_2S Mark IXA entered service in the Vickers Valiant, to be followed by the Avro Vulcan and the Handley Page Victor. The equipment was known overall as the NBS (Navigation and Bombing System), and comprised the H_2S Mark IXA integrated with the NBC (Navigation and Bombing Computer). I am indebted to Air Vice Marshal Hedgeland[7] for the following summary of this integrated system which employed doppler navigation originally known as *Green Satin* in addition to the Mark IXA radar.

---

† The Ashton aircraft had conventional aerodynamics as opposed to the delta and crescent wings of the V-bombers. Only two were built.

Green Satin was the first doppler navigation system to be developed by RRE. It used crossed slotted waveguide arrays looking *vertically downwards* which aligned themselves with the aircraft's track using doppler shift returns, so determining the drift angle, and the system determined ground speed accurately by comparing doppler shift between the forward and rearward returns. Green Satin information was fed into a ground position indicator (GPI) which presented Latitude and Longitude, initial settings of the counters being from visual or NBS fixes. In the standard installation there was no automatic intercommunication between Green Satin and NBS. The NBS fed airspeed from the air mileage unit and heading from the distant reading compass (DRC) into the NBC which computed the N/S and E/W airspeed components. To these were added windspeed components and the resultant ground speed components integrated to drive Latitude and Longitude counters on the NBC navigation panel and the shift circuits on the H_2S PPI, as in H_2S Mk IV/IVA. The windspeed components could be fed in manually to NBC initially and then corrected by using wind component shift voltages to stabilise the H_2S picture, the wind so determined being fed back into NBC. Fixes on accurately determined returns could also be fed into NBC from H_2S to correct the Latitude and Longitude counters. The link between H_2S and NBC was a highly complex electromechanical control unit (Type 585) produced by EMI.

The scanner of Mark IXA was a 6 foot long reflector fed by a slotted waveguide from a magnetron giving a pulsed output of between 100 and 200 kilowatts. The scanner rotated at either 8, 16 or 32 revolutions per minute and was roll stabilised to ± 30 degrees and pitch stabilised to ± 15 degrees. The vertical section of the reflector was shaped to produce a $cosec^2$ pattern and the azimuth beam-width was 1.5 degrees. Since the scanner was built into the chin position of the aircraft the rearward view was restricted. The rearward obscuration varied from about 60 to 90 degrees in the three types of V-bombers but was not operationally significant.

For the Valiants of 543 squadron at Wyton a modified version of H_2S Mark IXA was produced for reconnaisance. In this version the scanner could be locked in a sideways position using the aircraft's motion to scan a swathe of ground abeam, so becoming a SLAR (Sideways-Looking Airborne Radar). The drift angle derived from the Green Satin doppler system was used to align the scanner at right angles to the aircraft's track (as distinct from the aircraft's heading). The received signals intensity modulated a single cathode-ray tube trace across which a film was drawn at a rate proportional to ground speed derived from the Green Satin doppler system. With appropriate scaling of the trace and the rate of film transport, the film was exposed to produce a continuous radar map of the territory being passed. This film was developed by a 'rapid processor' produced by Kelvin and Hughes so that the results could be viewed in flight after a short delay. The equipment was known as 'Red Neck'.

In 1950 the Air Staff requirement was for a blind bombing accuracy of 200 yards with the bomber flying at 500 knots and at an altitude of 50 000 feet.

In the light of the estimated performance of the projected H_2S this requirement was discussed in a TRE memorandum in November 1950[8] and this led to the investigation of the use of H_2S on Q-band (8 millimetres). An experimental Q-band H_2S was constructed in TRE in 1951, but in the event the 3 centimetre Mark IXA appears to have performed well in the V-bombers until they were retired after more than 25 years service.

H_2S Mark IXA in the V-bombers was twice used in war. In 1956 Valiants used the system to make the pre-emptive strike against the Egyptian Air Force at Cairo airport. On 1 May 1982 a Vulcan made a 16 hour, 7000 mile round trip from Ascension Island and used the H_2S and NBS system in the pre-dawn attack with 21 1000 pound bombs[9] on the Port Stanley airport. That was the last use of H_2S successively modified from the original experimental system which first became airborne in the Halifax 40 years earlier.

References

1   Lovell TRE record p 49
2   *The history of TRE post design services at home and abroad 1942–45* part III pp 183–184 PRO file AIR 20/1532.
3   Bowen E G 1987 *Radar Days* (Bristol: Adam Hilger) pp 192–193 and pp 200–201
4   Guerlac Henry E 1987 *Radar in World War II* (Tomash Publishers and American Institute of Physics) pp 283–285 and Chapter 21
5   Paper BC/S26829/Radar. Meeting to be held at Headquarters Bomber Command at 14.45 hours on Friday 9 February 1945 to discuss H_2S policy (inserted in Lovell TRE record p 51)
6   Carter C J 1946 *H_2S an airborne radar navigation and bombing aid* in *J. IEE* **93** pt IIIA fig 13 p 459
7   Hedgeland P M S/Lovell correspondence 1990–1991 deposited with the Lovell papers in the John Rylands University Library of Manchester.
8   TRE memorandum No 362 dated 27 November 1950
9   *Task Force* Time-Scan Publication

Chapter 33

Envoi—1991

This book has been written more than half a century since I became a member of the wartime research establishment and 45 years after I was released to return to re-start my university career. I have described that return and the subsequent developments that led to the establishment of Jodrell Bank and the construction of the large radio telescope in *Astronomer by Chance*. This is not the place to repeat any of that story but in writing about the war years I have been forcibly reminded of the dramatic effect of those years on the subsequent careers of those of us who were in our twenties when the war began and were wrenched from the calm of the university laboratories.

During the war years the scientists of the Allied powers had exerted a great influence on massive military operations. A new faith in the powers of science and scientists had arisen and for a decade or so after the end of the war this faith was transferred to peacetime developments. If an operational need for a scientific device was established, the question during the war was never 'How much will it cost', but 'Can you do it and how soon can we have it?'. The 'doing of it' involved us with guns, ships or aircraft and integrated us with industry and the disciplines of the military organisations.

These attitudes and influences had a fundamental effect in the years of peace. The faith in science was manifested in the encouragement from the Establishment to re-create the scientific endeavour and in this re-creation the attitude of the scientists was inevitably influenced by their wartime careers. Massive instruments for atomic and nuclear research appeared and the rockets of war were developed to place scientific instruments in orbit around the Earth. In my own case it had been easy to ask for a Lancaster bomber and build a protuberance under the fuselage to house a large aerial. The transfer to peace was the belief that if it was easy to fly a six foot aerial in a bomber aircraft at 20 000 feet then it should be possible to build one forty times the size on the ground. It was done, but it was not easy—as the bureaucratic organisations of peace grew in strength so the faith in scientists and their activities came under attack. Nevertheless the concepts of big science—nuclear machines, huge telescopes, earth satellites and space probes—had been established and a new generation of scientists were imbued with the relevant attitudes to scientific research. There can be no turning back.

Appendix 1

Note on Minelaying by H₂S

In the early stages of the war, mines were laid in open waters by aircraft of Coastal and Bomber Commands (mainly by 5 Group of Bomber Command). In a report to Churchill on 3 April 1942, Harris specifically identified the mining operations as an important contribution made by his Command. 'Extensive minefields have been maintained, and are now being greatly increased by the Command in the Bight, the Belts and the Baltic and the approaches to German ports and to the Western bases'.[1] His memorandum stated that from the beginning of the war to 31 December 1941, 2569 mines had been laid and that 1757 of these had been laid by Bomber Command and 812 by Coastal Command. At that time Harris stated that the known losses of enemy shipping caused by the mines was 211 360 tons. Hitherto the mines had been laid by Hampdens which could carry only one mine but the operation was in process of transference to the heavy bombers—Lancasters and Stirlings could carry six mines, and the Halifax and Wellington two. This led to a substantial increase in the number of mines laid—according to Saward[2] in the nine months from 1 January to 30 September 1943, Bomber Command laid 11 000 mines.

When H₂S became operational in the main bomber force, new techniques of mine laying were employed enabling the bombers to operate at altitudes of 15 000 feet or above. As Roskill[3] points out, this made the minelaying operation far less hazardous to the aircraft. This new method using H₂S Mark II was introduced in January 1944 and Roskill[4] gives the following statistics from that point to the end of the war (Home Waters only):

	Sorties	Mines laid	Tonnage of enemy shipping	
			Sunk	Damaged
1944 January–May	3221	9637	61 541	28 134
June–December	1910	7863	74 545	100 915
1945 January–May	991	4582	164 330	117 951

References

1 Saward Dudley 1984 *'Bomber' Harris* (London: Cassell) p 128
2 Saward Dudley 1984 *'Bomber' Harris* (London: Cassell) p 213
3 Roskill S W 1960 *The War at Sea 1939–1945* (London: HMSO) vol 3 pt 1 p 288
4 Roskill S W 1960 *The War at Sea 1939–1945* (London: HMSO) vol 3 pt 2 p 142 and p 235

Appendix 2

Glossary and Abbreviations

ACAS	Assistant Chief of Air Staff (three posts for Operations, Policy and Intelligence)
ADEE	Air Defence Experimental Establishment. In 1942 this became RRDE—the Radar Research and Development Establishment, and eventually ADRDE—the Air Defence Research and Development Establishment. This establishment was amalgamated with TRE in the early 1950's to form RRE—the Royal Radar Establishment. In the 1970's with further amalgamation RSRE was established (see TRE).
ADRDE	See under ADEE
AI	Air-Interception Radar, the radar equipment in night fighters used for the detection of enemy aircraft. AI Marks I to IV worked on a wavelength of 1½ metre and were superseded by the centimetre AI systems (qv).
AIF and AISF	The code used in TRE to describe a radar in a night fighter that could 'lock' the antenna onto the radar echo from the target and ultimately lead to 'blind firing'—that is, destruction of the target without visual contact. The production version was known as AI Mark IX.
AIH and AIS	See under centimetre AI
AM	Air Ministry
AMRE	Air Ministry Research Establishment (see TRE)
AMU	Air mileage unit
AOC	See C-in-C and CO
ASE	Admiralty Signals Establishment (HM Signal School until August 1941)
ASV	Air to Surface Vessel, the radar equipment in aircraft used for the detection of ships and surfaced submarines. ASV Mark I and II worked on a wavelength of 1½ metre. ASV Mark III was the 10 centimetre adaptation of H_2S. ASV Mark VI was a high power 10 centi-

265

	metre version with attenuator and ASV Mark VIA was the lock–follow version. ASV Mark VII was the British 3 centimetre version.
ASVS	The original 10 centimetre ASV system under development in TRE. Abandoned in the autumn of 1942 in favour of the adaptation of H₂S to ASV.
Azimuth	The position around the horizon. In the Northern Hemisphere the convention is that due north is 360 degrees (or zero), east 90 degrees, south 180 degrees, and west 270 degrees.
BDU	Bomber (or Bombing) Development Unit
BRS	Bawdsey Research Station (see TRE)
CAS	Chief of the Air Staff
Centimetre AI	A radar airborne interception system for night fighters that was developed in the early years of World War II and in which the wavelength of the radar was generally 10 centimetres or less. This enabled a narrow beam of radar waves to be transmitted from the aircraft and thereby avoid the confusing ground returns (qv). The development forms in TRE were known as AIS (spiral scanning) or AIH (helical scanning). The crash programme version of AIS became AI Mark VII and the main production AI Mark VIII. The American 10 centimetre AI (SCR 720) with a helical scanner coded as AI Mark X eventually superseded AI Mark VIII in the RAF night fighters.
CC	Controller of Communications in the Air Ministry (see CTE)
CCE	Controller of Communications Equipment in MAP (see CTE)
CH	Chain Home, the network of radar stations built along the east and south coasts of England for the detection of enemy bombers. It played a vital role in the Battle of Britain.
CHL	Chain Home Low. A radar system added as a supplementary radar on the CH stations to give warning of low flying aircraft.
C-in-C	Commander in Chief. For an officer of Air Rank (Air Commodore or above) the formal title is Air Officer Commanding-in-Chief—colloquially the AOC.
CO	Commanding Officer of a unit (Station or Squadron). If of Air Rank (Air Commodore or above) the formal title is Air Officer Commanding ... (see also C-in-C).
CRT	Cathode-ray tube

CTE	Controller of Telecommunications Equipment: post occupied by Sir Frank Smith. In mid-1942 this post was superseded by the creation of the post of CCE (Controller of Communications Equipment in MAP) and of CC (Controller of Communications in the Air Ministry). Sir Robert Renwick occupied both these posts.
CVD	Committee for the Co-ordination of Valve Development.
D B Ops	Director of Bomber Operations
DCAS	Deputy Chief of Air Staff
DCD	Director of Communications Development (in MAP)—post successively occupied by Sir Robert Watson-Watt, Sir George Lee and Air Commodore Leedham.
DGS	Director General of Signals (Air Ministry)
Dipole	A rod-like antenna used for transmitting and receiving radio waves. A common form is known as the half-wave dipole, in which a rod one-half the length of the wavelength is split at the centre point and connected by cable to the transmitter or receiver. This type of antenna has a broad beam of reception. (See also Yagi aerial.)
D of Ops Tac	Director of Operations Tactical Air Force
D of Radar	Director of Radar (Air Ministry)
DRP	Director of Radio Production (in MAP)
Fishpond	Tail warning system in RAF bombers working on a wavelength of 10 centimetres (see also Monica)
FIU	Fighter Interception Unit
GCI	Ground Controlled Interception
Gee navigation aid	A navigational and bombing aid first used operationally by the Royal Air Force over Germany in March 1942. Three widely separated transmitters in England transmitted pulsed radar signals simultaneously. The centre station acted as the 'master' and the other two as 'slaves'. A receiver in the aircraft measured the difference in time of receipt of the pulses from each slave and the master and thereby determined its distance from the master and each slave. The locus of all points at which a constant time difference is observed between the pulse from the master and one of the slaves is a hyperbola. Special charts in the aircraft of these sets of intersecting hyperbolae enabled the navigator to determine the position of the aircraft. The range is

limited by the earth's curvature because the aircraft receives the direct ray from the transmitter. With the transmitters on the east coast of England, the maximum range of about 350 miles encompassed the Ruhr. At this range the position of the aircraft could be calculated to within an elliptical area of about 6 miles by 1 mile.

GL   Gun laying radar used by the Army to direct the anti-aircraft guns.

Ground returns   The signals scattered from the ground by a radar in an aircraft. In the early years of the war the radar in the night fighters used antennae transmitting in a broad beam. The signals scattered from the ground obscured the signal scattered from the target and thus limited the useful range to the height at which the fighter was flying. The development of narrow-beam centimetre equipment overcame this difficulty, and in 1942 it was found that a narrow radar beam on a wavelength of 10 centimetres or less gave enough discrimination in the ground returns to produce a useful map. This investigation led to the development of the H_2S bombing aid.

H_2D A pulsed doppler version of H_2S for the detection of moving vehicles.

H_2S The centimetre navigational and blind-bombing radar system developed for use by the night bombers of the Royal Air Force in World War II. See Appendix 4 for the various versions.

H_2X The American version of the British 3 centimetre H_2S system

HQBC   Headquarters of Bomber Command (at High Wycombe)

IFF   A responder device in aircraft for the identification of friend or foe

K-band   Wartime usage referred to a wavelength of 1.25 centimetres

Klystron   A form of vacuum tube developed in the United States in the 1930s for the generation and amplification of microwaves. The klystron was widely used when radars working on wavelengths of 10 centimetres or less were developed in World War II. A beam of electrons is 'bunched' by means of an alternating voltage applied to grids in the tube. A system of resonant cavities enables microwave power to be obtained with a frequency equal to that of the bunching.

Leigh light	In the early years of World War II the German U-boats tended to remain submerged during the daytime but to surface at night to recharge their batteries. The development of ASV radars enabled the night flying aircraft of the RAF Coastal Command to detect the U-boats in darkness, but, except under occasional conditions of favourable moonlight and good visibility, they could not carry out attacks, because the U-boat could not be positively identified. In 1942 a powerful searchlight with a flat-topped beam was installed underneath the fuselage of the radar-equipped Coastal Command night patrols. The searchlight was mounted in a retractable cupola, and during the final approach to a target detected by radar it was lowered and the target illuminated so that a visual identification could be made. The device was known as the Leigh light, after its inventor, Wing Commander H Leigh of RAF Coastal Command.
Magnetron	The special form of magnetron that became very important in many radar devices used in World War II is known as the *cavity magnetron*. The underlying principle was discovered by J T Randall and H A Boot in the University of Birmingham and was quickly applied to the development of a powerful generator of microwaves in the centimetre wavelength region. The cavity magnetron consists of a copper block in which cylindrical cavities are drilled. Electrons produced from a cathode in the centre of the block are accelerated toward the cavities, but a magnetic field aligned with the axis of the block forces the electrons into spiral paths. The electromagnetic fields from the successive cavities bunch the stream of electrons, and oscillations are stimulated in the cavities, which can be extracted to form a powerful microwave signal.
Magslip	In the application described in this book the magslip was mounted on the scanner, with the rotor geared to the shaft of the rotating scanner. The stator had two windings through which waveforms were generated 90 degrees out of phase and connected to the PPI (qv) so that the time base rotated in synchronism with the scanner.
MAP	Ministry of Aircraft Production
MAPRE	Ministry of Aircraft Production Research Establishment (see TRE)

Metox	A listening receiver in the U-boats to detect the approach of a Coastal Command aircraft using 1½ metre ASV (see also Naxos)
MIT Radiation Laboratory	See Radiation Laboratory
Monica	Tail warning system in RAF bombers working on a wavelength of 1½ metre (see also Fishpond)
Naxos	A listening receiver in the U-boat to detect the approach of Coastal Command aircraft using 10 centimetre ASV (see also Metox). Also used in German night fighters to home in on the H_2S transmissions from RAF bomber aircraft.
NBC	Navigation and Bombing Computer
NBS	Navigation and Bombing System
Oboe	An accurate bombing aid first used operationally by the Royal Air Force in December 1942. Two radar stations on accurately surveyed positions on the east coast of England controlled an aircraft that carried a beacon transmitter so that it could be guided to fly at a constant distance from one of the ground stations (the 'cat' station). Thus the aircraft flew in a circle with the cat station at its centre. The radius of the circle was chosen so that the aircraft flew over the target. The other ground station (the 'mouse') also measured the range from the beacon carried in the aircraft and transmitted the 'release' signal to the aircraft at the calculated time. As in the case of *Gee*, the range was limited because the aircraft had to observe the direct radar transmission from the ground stations. Although the accuracy was very high—at a range of 250 miles an aircraft flying at 30 000 feet could release its bombs to within 120 yards of a selected spot—only one aircraft could be controlled every ten minutes. However, the use of high-flying Mosquito aircraft fitted with *Oboe* to mark the target by dropping flares led to great devastation in the Ruhr.
ORS	Operational Research Section (of Bomber Command)
PDS	Post Design Services—a TRE organisation for assisting with the introduction of radar systems to operational use.
PFF	Pathfinder Force (8 Group) of Bomber Command
Polar diagram	The shape of the beam of radiation emitted (or received—the two are identical). The beam width is usually defined as the width of the beam when the intensity is one-half the maximum. There are alternative methods

of picturing the shape of the beam. In the conventional polar diagram the antenna is the 'pole', or origin, of the beam. The radiated pattern normally consists of a prominent forward lobe with a number of (unwanted) minor or side lobes.

PPI	Plan position indicator (on a cathode-ray tube) originally known as radial time base (RTB)
PRF	Pulse repetition frequency
Radiation Laboratory	The organisation for the development of radar in the USA set up in the Massachusetts Institute of Technology (MIT) in the autumn of 1940
RDF	Radio direction finding—the early code for radar
RF	Radio frequency
RPU	Radar Production Unit. A special organisation created in association with TRE for the manufacture of radar sets.
RRDE	See under ADEE
RSRE	See TRE
RTB	See PPI
S-band	The 10 centimetre band of wavelengths. (Most of the S-band H_2S and ASV systems worked on a wavelength between 9 and 10 centimetres.)
Schnorkel	A breather tube fitted to the U-boats enabling the main hull to remain submerged whilst charging their batteries.
TR (box or system)	Common transmit–receive system. A device to enable the same aerial to be used both for transmitting and receiving.
TRE	The British wartime radar establishment. The original pre-war radar research department was at Bawdsey Manor, on the East coast of England, and was known as the Bawdsey Research Station (BRS). When war broke out, the establishment moved to Dundee, in Scotland, and became the Air Ministry Research Establishment (AMRE). In the spring of 1940 this establishment moved to the Dorset coast in the south of England. With a change in ministerial arrangements, AMRE became MAPRE (the Ministry of Aircraft Production Research Establishment); finally, in November 1940, it became TRE (Telecommunications Research Establishment). In the spring of 1942 TRE was evacuated to Great Malvern, in Worcestershire, and in the post-war years various mergers and further changes of name took place. It is now (1991) the Royal Signals and

	Radar Establishment (RSRE)—see also under ADEE.
u/s	Unserviceable
VCAS	Vice-Chief of the Air Staff
VCCE	Vice-Controller of Communications Equipment in MAP. When Renwick was appointed CCE (qv) and Controller of Communications in the Air Ministry in the summer of 1942, Watson-Watt became VCCE and SAT (Scientific Adviser on Telecommunications) in the Air Ministry.
Window	Tinfoil strips dropped from RAF bombers to confuse the German radars
X-band	The 3 centimetre band of wavelengths
Yagi aerial	A device for obtaining a narrow beam for the reception or transmission of radio waves, developed by the Japanese engineer Hidetsugu Yagi in 1928. This consists of a half-wave dipole usually backed by a single rod acting as a parasitic reflector and a number of rods in front of the dipole acting as parasitic directors. This type of directional antenna has been widely used, mounted on housetops for television receivers.

Appendix 3

Principal Staff and Command Appointments

referred to in this book during the main phase of H₂S/ASV

TRE

Chief Superintendent *A P Rowe*

Deputy Chief Superintendent *W B Lewis*

Superintendent 3 *P I Dee*

MAP (Ministry of Aircraft Production)

CCE (Controller of Communications Equipment) *Sir Robert Renwick*

VCCE (Vice Controller of Communications Equipment) *Sir Robert Watson-Watt*

DCD (Director of Communications Development) *Air Commodore Leedham*

AM (Air Ministry)

S of S (Secretary of State for Air) *Sir Archibald Sinclair*

CAS (Chief of the Air Staff) *Marshal of the Royal Air Force Sir Charles Portal*

VCAS (Vice-Chief of the Air Staff) *Air Chief Marshal Sir Douglas Evill*

DCAS (Deputy Chief of the Air Staff) *Air Marshal Sir Norman Bottomley*

ACAS (Ops) (Assistant Chief of the Air Staff (Operations)) *Air Vice-Marshal W A Coryton*

D B Ops (Director of Bomber Operations) *Air Commodore S O Bufton*

CC (Controller of Communications) *Sir Robert Renwick* (also CCE)

SAT (Scientific Adviser on Telecommunications) *Sir Robert Watson-Watt* (also VCCE)

DGS (Director General of Signals) *Air Marshal Sir Victor Tait*

Bomber Command

C-in-C (Air Officer Commanding-in-Chief) *Air Chief Marshal Sir Arthur Harris*

Deputy C-in-C (Deputy Air Officer Commanding-in-Chief) *Air Marshal Sir Robert Saundby*

SASO (Senior Air Staff Officer) *Air Vice-Marshal H S P Walmsley*

Chief Radar Officer *Group Captain Dudley Saward*

CO 5 Group (Air Officer Commanding 5 Group) *Air Vice-Marshal The Hon R A Cochrane*

CO 8 Group (Air Officer Commanding 8 Group (Pathfinder Force)) *Air Vice-Marshal D C T Bennett*

Coastal Command

C-in-C (Air Officer Commanding-in-Chief) *Air Marshal Sir Philip Joubert de la Ferté* and *Air Marshal Sir John Slessor* (from 1943)

Appendix 4

Summary of H₂S Systems

Mark

I	The first 'crash programme' 10 centimetre version for the Pathfinder Force.
II	The main production 10 centimetre system for the main force.
IIA	Horn feed to scanner instead of dipole.
IIB	Mark II + Fishpond
IIC	Mark II + scan-corrected PPI (Type 184) + roll stabilised and improved scanner
III	3 centimetre, six operational December 1943 in PFF.
IIIB (interim stage between III & IIIA)	Mark III + scan-corrected PPI (Type 184)
IIIA	Mark IIIB + roll stabilised scanner
IIID	Mark IIIA + higher powered magnetron
IIIC (whirligig)	IIIA + 6 foot scanner
IIIE	Improved IIIA (magnetic tube indicator Type 216, barrel scanner, shorter pulse length)
IIIF (whirligig)	IIIE with 6 foot scanner
IIIG	Conversion of 10 centimetre IIC system to 3 centimetre operation (non-stabilised scanner) modifications for long-range detection of coast lines and operations in Pacific
IIIH	as IIIG with electromagnetic indicator Type 216
IV	3 centimetre as for IIIA with computing and wind finding, range stabilisation and automatic scan-

275

	correction with altitude. Linkage with Mark XIV bombsight
IVA	As IV + 6 foot scanner linked with AMU and Mark XIV bombsight + ground stabilised display
V	This Mark of H_2S does not appear in the sequence of British versions described in this book. It is believed to have been reserved for the American 3 centimetre H_2X but this was never introduced into RAF Bomber Command.
VI	1¼ centimetre with 6 foot scanner (in Lancasters) or 28 inch scanner (in Mosquitoes)
VII (Lion Tamer)	Fully developed 1¼ centimetre H_2S with computing facilities linked to Mark XIV bombsight
VIII	Concept of Mark VII integrated with NBC computer using 3 centimetres
IX/IXA	Final versions of 3 centimetre H_2S with 6 foot scanner integrated with design of V-bombers and with aircraft navigation and bombing computers

Note

After VE day all Marks of H_2S to IIIG were declared obsolescent. The K-band Mark VI and VII were subsequently abandoned. Mark VIII was a development version. Mark IV was superseded by IVA before production commenced. Mark IVA was the standard post-war H_2S until superseded by H_2S Mark IXA in the V-bombers in 1956.

Index

'Acorn tubes', 30
ADEE, *see* Air Defence Experimental Establishment
Admiral Graf Spee, 22
Admiralty, 121, 170
Admiralty Signals Establishment (ASE) (HM signal school), 51–3, 61, 171
Admiralty Valve Laboratory, 31, 181
ADRDE, *see* Air Defence Research & Development Establishment
Aerial towers, Staxton Wold, 6–7
Aeronautical Research Committee, 1
AGLT, *see* Air Gun Layer Turret
Airborne interception radar (AI), vii, ix, 12–14, 16–17, 21–6, 29–30, 36, 41, 45–6, 50–1, 54, 101
    first centimetre, 56–7
    lock–follow (AIF, AISF), 69–84
    Mark VII, 65–7, 100, 105, 107, 142
    Mark VIII, 65–7
    Mark X, *see* SCR720
Air to Surface Vessel (ASV), viii, ix, 19, 22–6, 84, 146, 155–70
    see also Centimetre ASV
Air Council Room, 214–5
Aircraft, *see specific types of*
Aircraft & Armament Experimental Establishment (AAEE), 99
Air Defence Department (RAE), 29
Air Defence Experimental Establishment (ADEE), 48
Air Defence Research & Development Establishment (ADRDE), 49, 240
Airey, Joe, 27
Air Gun Layer Turret (AGLT), 207
Air mileage unit (AMU), 219–20
Air Ministry, 1–2, 4–5, 10, 12, 20, 22, 30, 80, 100–1, 105–6, 135, 139, 158, 168, 184, 201, 208, 243
Air Ministry Research Establishment (AMRE), 10, 15, 21, 25–7, 29, 37, 42–3, 49, 51, 55
Air position indicator (API), 219–20
Airspeed indicator (ASI), 77
Air Staff, 85–7, 108, 153
AIS, 47–8, 54
    flight trials, 63–5
    Mark VII/Mark VIII, 55–68
    see also Airborne interception radar
AISF (AIF), *see* Airborne interception radar
Alamein, 133
Alcock, N Z, 186, 226
Alconbury (USAF station), 195
Algiers, 172
American 8th Air Force, 148
American 8th Bomber Command, 193–6
AMRE, *see* Air Ministry Research Establishment
Anderson, Sir John, 122
Anglo–Australian telescope, 11
Anson aircraft, 24, 89
Anti U-boat Committee, 167
    see also U-boats
Army, and H₂S, 238–45
Ascension Island, 260
ASE, *see* Admiralty Signals Establishment
Ashton aircraft, 258
ASI, *see* Airspeed indicator
ASV, *see* Air to Surface Vessel
Athenia, 8
Atkinson, J R, 29, 38–9, 42, 45, 50, 59, 67, 181–2

Barkhausen–Kurz effect, 33

Barry (S. Wales), 20, 24, 28
Batt, Reg, 41–2
Battle aircraft, 12–13, 64, 79
Battle of the Atlantic, 85, 155–70, 246–50
Battle of Britain, vii, 5
Bawdsey Manor, 4–5, 7, 12, 21, 70
Baxter III, James Phinney, 26
Bay of Biscay, viii, ix, 156, 161–3, 166, 169, 246
BD project, 24
BDU, *see* Bomber Development Unit
Beattie, R K, 21–4, 155
Beaufighter aircraft, 15, 55–6, 65–7, 74, 76, 78–81, 85, 95
    X7579, 65
Beaverbrook, Lord, 87, 121
Beeching, I, 159, 182
Belgium, 110
    invasion of, 29
Bembridge (Isle of Wight), 172
Bennett, D C T (Don), 144, 152, 175, 181, 188, 215–6, 221, 223, 227, 229, 243, 251–2
    biography of, 137–9
Bentley Priory, 5
Berlin, 178, 180–93, 211–2, 234, 253
Bernal, J D, 104–5
Biggin Hill, 5
Birmingham (bombing of), 104
Birmingham University, vii, 31, 35–7, 48, 61, 181
Blackett, P M S, 1–4, 8, 10–11, 17–18, 21, 23–4, 70–1, 75, 104–5, 169–70
Black Mountains, 95
Black Sea, 151
Blenheim aircraft, 12–16, 21, 28, 39, 41, 43, 50, 55, 59, 62–3, 71, 77–9, 95, 103–4, 126, 150
    N3522, 57, 63–4, 157
    T1939, 76–8
    V6000, 91–4, 97, 99–102, 105–7, 142
Blind bombing, 11
Blind navigation (BN), 85, 97
Blohm shipyard, 176
Blumenau, R, 121, 123

Blumlein, A D, 15, 71, 107, 127, 142
Blythen, F, 127, 142
BN, *see* Blind navigation
Boeing aircraft, 181–2
Bomber Command, ix, 80–1, 85–6, 88, 96, 99, 101–2, 104, 108, 127, 136–7, 142–4, 153, 159–61, 165, 168–9, 173, 175, 178, 182–3, 203, 206, 210, 217, 220, 225, 233–5, 238, 251
    and H2S problems, 197–205
    conflict with, 210–6
    HQ (High Wycombe), 101–2
Bomber Development Unit (BDU), 106, 143, 148–9, 152, 185
Bond, W M, 76
Bonner, 183–4
Boot, A H, vii, 33, 35
Boscombe Down, 99
Bottomley, Air Marshal Sir Norman, 168, 214–5
Boulogne, 239
Bowden, Lord (of Chesterfield), 121
Bowden, Vivian, *see* Bowden, Lord (of Chesterfield)
Bowen, E G ('Taffy'), vii, 3–4, 7, 10, 12, 14, 16–18, 20–3, 25–7, 30, 49, 58, 88–9, 103, 145–8, 155
Bragg, W L, 4
Bremen, 252
Bristol, 54, 112
Bristol Aircraft Co, 65, 76
Bristol Channel, 23, 93
Bristol University, 31, 61
Brown, Commander J D, 83
Browne, C O, 127, 142
Bruneval raid, 121
BTH (Rugby), 48–50
Burcham, Professor W E, 29, 32, 37–8, 41, 50–2, 59, 65–7, 82, 92, 96, 117, 124
Butler, R A B, 122
Butt, D M, 86
B24 Liberator, 146
    see also Liberator aircraft

Cairo, 151, 260

Calcutt, Corporal, 103, 160

Callendars Cable & Construction Co, 48

Callick, E B, 37, 58

Cap d'Antifer, 239

Carter, C J, 101

Casablanca Conference, 173, 193

'Cat' stations (Oboe), 88

Cathode-ray tube (CRT), 8, 14, 42, 60, 64, 77, 93, 95, 99, 208, 221, 225

Cavendish Laboratory (Cambridge), 4–5, 29, 37, 59

Cavity magnetron, *see* Magnetron

Centimetre ASV, 28–9
  and Naxos, 165–70
  and U-boats, 155–64
  development of, 157–8
  see also Airborne radar

CH, *see* Chain Home radar stations *and also* Chain Home Low radar installations

Chadwick, Roy, 149, 222

Chain Home Low (CHL) radar installations, 7, 14, 42

Chain Home (CH) radar stations, 3–7, 10–11, 13–14, 42

Chamberlain, Group Captain, 16, 159

Chamberlain, Neville, vii, viii
  resignation of, 29

Chapman, A H, 24–5

Cheltenham, 144–5

Cherbourg, 119–20, 172

Cherry, E C, 57

Cherwell, Lord (F A Lindemann), 2, 86, 91, 94, 97, 100, 104–5, 127, 133–6, 183–4, 188, 199, 222, 224

Chesterman, Joyce (Mrs B Lovell), 4, 23–4, 110–8

Chisholm, 172

Chivenor (RAF station), 160–2

CHL, *see* Chain Home Low installations

Christchurch (aerodrome), 27–8, 47, 57–8, 65–6, 76–7, 80

Churchill, Winston, 2, 29, 39–40, 42, 85–7, 100, 104, 111, 127–8, 132–6, 139, 148, 150–1, 155–6, 172, 239, 247

C-in-C Bomber Command, 85
  see also Harris, A T

C-in-C Coastal Command, 159
  see also Joubert de la Ferté *and* Slessor, J

C-in-C Fighter Command, 12, 15, 18, 41

Civilians, during World War II, 100–18
  food queues, 119–25

Clarendon Laboratory (Oxford), 63

Clayton, Sir Robert J, 45, 57

Coastal Command, viii, 22, 155–6, 158–64, 166, 170, 211, 233, 263

'Cobweb', 172

Cochrane, C A, 53

Cochrane, Vice Marshal The Hon R A, 144, 215–6, 251

Cockburn, R, 231, 235, 238

Cockcroft, J D, 4–5, 11

Code name H$_2$S, 96–7

Cole, E K (Malmesbury), 66

Cologne, 132, 153

Coltishall (RAF station), 80–1

Columbia Radiation Laboratories, 224

Combined Operations, 171–2

Common transmit–receiver system, 62–3

Communications Development, Director of (DCD), 5, 24–5, 28, 46, 143, 157, 203, 240

Coningham, Air Marshal Sir A, 241

Controller of Telecommunications Equipment (CTE), 122
  see also Smith, Sir Frank *and* Renwick, Sir Robert

Cooke, A H, 63

Cooke-Yarborough, E H, 15

Corfe Gap (Purbeck peninsula), 120

Cossor, A C, 7, 12

Croney, J R, 53

Crossed dipoles, 6–7

CRT, *see* Cathode-ray tube

CTE, *see* Controller of Telecommunications Equipment

Cunningham, Group Captain, 56

Curran, Jane (*née* Strothers), 29

Curran, S C, 29, 50, 59–60, 66–7, 107

CW magnetron, 39

Dale, Corporal, 103
Dalton-Morris, Air Commodore, 234–5
Daventry transmitter, 3
Day, Flight Lieutenant J H, 224
DCD, *see* Communications
    Development (Director of)
D-Day, viii, 49, 172, 203, 238–45
Dee, Professor P I, 29, 37–8, 41–2,
    45–6, 48–53, 56–7, 65–6, 70,
    72, 78, 82, 85, 90–4, 97, 99,
    103, 117, 122, 124, 132–3, 141,
    146–7, 153, 157, 168, 171,
    180–1, 183–4, 195, 214, 221–2,
    235, 240, 247, 252
Defford aerodrome, 126–8, 130, 135,
    137, 139, 142, 158–9, 167,
    194–5, 207, 217, 224–5, 239,
    257–8
Denmark, invasion of, 27
Department of Scientific & Industrial
    Research (DSIR), 2, 5, 10
Deutschland, 22
Devons, Sam, 29, 221, 224
Dickens, B G, 235
Dickie, Flight Lieutenant John, 97,
    102, 106, 199
Dipoles, 6–7, 13, 174
Dippy, R J, 87
Display stabilization, 219
Distant reading compass, 259
Don, the river, 151
Donaldson, Squadron Leader O R,
    106, 229
Dönitz, Admiral, 156, 166, 248
'Doorknob' tubes, 30
Doppler (navigation), 240–1, 258–9
Double paraboloid helical scanner, *see*
    Paraboloids
Dowding, Sir Hugh, 41
    see also C-in-C Fighter Command
Downing, A C, 57, 80–1, 123
DSIR, *see* Department of Scientific &
    Industrial Research
Du Bridge, Lee, 147, 183–4
Duddell medal (awarded to J T
    Randall), 33

Duke, S M, 36
Dundee, 10, 12, 14, 17–21, 24–5,
    27–8, 39, 55, 65, 70
Dunkirk, 39, 110
Durnford House, 50

Eaker, Brigadier General Ira, 173,
    193–6
Edwards, George, 57, 63, 65
Egypt, 132–3, 260
Eisenhower, General, 203
Elizabeth, Queen Elizabeth, 24, 117–8
Emden (bombing of), 195
Emergency Service Organisation, 121
EMI, 15, 30, 71, 106, 126–7, 133–4,
    136, 148, 150, 157, 171, 199,
    217, 258
    and Halifax R9490, 107–8
Emmett, Flight Lieutenant Peter, 6
English Electric E3/45, 258
E series cavity magnetrons,
    E1188 sealed-off magnetron, 51,
        58–9
    E1189 oxide-coated magnetron, 58
    E1198, 51, 58–9
Espley, D C, 45, 57, 62
Essen (bombing of), 132
Excellent, 82

Farnborough, 24, 27
    see also Royal Aircraft
        Establishment
Feather, W, 5
Ferranti, 72–4, 157–8
Fertel, G E F, 5
Fighter Command, 11, 14–16, 80
Fighter Interception Unit (FIU), 16,
    65, 82–3, 241
Fishpond tests, 206–10, 236–7, 252
FIU, *see* Fighter Interception Unit
Fleet Air Arm, 22
'Flensberg', 236
Forres (school), 50, 103
Fortescue, Richard, 161
Fortress aircraft, viii, 193, 195–6
France, 111, 156, 168–9, 193, 246
    invasion of, 39
Frankfurt (bombing of), 195

Freeman, Captain Spencer, 121–2
Fry, D W, 158

Gaunt, H C A, 121
Gavin, M R, 25
GCI, *see* Ground controlled interception
GEC, 24–5, 30–1, 33, 36, 45–6, 51, 56, 58–9, 62, 65–7, 69, 91, 93
Gee navigational system, 87–8, 101–2, 106, 193, 200, 206
Geiger counter, 59–60
George VI, King, 24, 117
German Air Force, *see* Luftwaffe
German–Soviet pact, 6
Germany, vii, viii, 8, 22, 105, 127, 146
GH (navigational aid), 238
Gilbert, C W, 29
Gilfillan, Squadron Leader, 172
GL, *see* Gun-Laying equipment
GL3 (gun-laying radar), 49
Gladiator aircraft, 79
Gloucester, 144–5
Goebbels, Paul Josef, 188
Gollin, E M, 171–2
Gomorrah, Project, 175
Goniometer coil, 6
Gramophone Co, 158, 202, 220, 258
 see also EMI
Graveley (RAF station), 150, 152
Great Malvern, 116–7, 119–25, 240
Green Satin (doppler navigation), 258–9
Griffiths, J H E, 63
Ground controlled interception (GCI), 7, 14, 55
Ground position indicator, 259
Guerlac, Henry E, 49, 145
Guldborg Sound, 189
Gun-Laying (GL) equipment, 45, 47–50, 62, 69, 158
Gutton, H, 36
G22 camera gun, 79
G45 cine gun, 79

Halifax bombers, 126–31, 142, 154, 238, 263
 BB360, 208
 HX166, 240–1

R9490, 133, 136, 144
 and EMI, 107–8
V9977, 99–109, 135, 139, 198–9
 crash of, 128–30
W7711, 135, 144
W7808, 143, 148
Hallade recorder, 76
Hamburg (bombing of), viii, 172–80
Hampden aircraft, 99
Hanbury Brown, R, 14, 16, 89
Handley Page works, 99, 100, 106–7, 149
Hanover (bombing of), 178
Hansen, W W, 31–2, 34
Harris, Air Marshal Sir Arthur T, 99, 134, 138, 144, 173, 181, 194, 203, 208, 213–4, 245, 263
Hartley, Sir Harold, 76
Hartree, D R, 70
Harvey-Bailey, A, 129
Hawkeye, Project, 248
Hayes Laboratory, *see* EMI
H_2D, 240–1
Hedgeland, Air Vice-Marshal P M S, 253, 255–7
Heinkel bomber, 40, 42
Hensby, G S, 91–3, 95, 99, 106, 123, 126–7, 143
Hertz, 34
Hewitt, Sir Ludlow, 49
High-frequency generation, 30–1
Hill, A V, 1–2
Hillman, Flight Lieutenant Peter, 143, 173, 186, 194, 254
Hinckley, Corporal, 103, 172, 254
Hitler, Adolf, 8, 27, 29, 110, 151, 155, 162, 166, 178, 249
HMS ships, *see under specific names of ships*
HM signal school, 51, 61
 see also Admiralty Signals Establishment
Hodgkin, Alan, 24, 25, 30, 46–7, 50, 52, 63–5, 67, 69, 72, 124, 207
Hodgkin/Nash & Thompson spiral scanner, 56–7
Holland, 110, 152, 233
 invasion of, 29

Holt Smith, C, 158, 231
Home Guard, 119
'Home Sweet Home', 97
Horton, C E, 51–2
Howse, Commander Derek, 53, 83
H₂S, vii–ix, 19, 60, 99, 105–8, 126–7,
    132–4, 136, 139, 141–3, 151–2,
    165–6, 172–8, 251–60
    Army and, 238–45
    background to, 85–98
    Bomber Command problems,
        197–205
    code name, 96–7
    conflict with Bomber Command,
        210–6
    Fishpond tests, 206–10
    K-band, 222–3
    Lion Tamer (Mark VII), 224–5,
        257
    main systems (1944), 204
    Mark II, 201–4
    Mark III, 201–4
    Mark IIIC, IIIE, IIIF, 225–7
    Mark IIIG, 256
    Mark IV/IVA, 217–20, 256–7
    Mark VI, 224, 227, 230–1, 257
    Mark VII, 224–5, 257
    Mark VIII, 257–8
    Mark IX/IXA, 258–60
    minelaying, 263–4
    and Naxos, 233–7
    new versions, 217–28
    opposition to, 143–8
    on 3 centimetres (X-band), 180–93,
        220–3
    and Pathfinder Force, 148–50
    on tank landing craft, 171–2
    and the USA, 183–4
    US 8th Bomber Command and,
        193–6
    Whirligig, 225–7
    X-band, 180–93, 220–3
Hull (bombing of), 104
Hurn aerodrome, 28, 85, 100–1, 103,
    107–8, 119, 126, 130
Hurricane aircraft, 5, 78
Huxley, Andrew, 75
Huxley, L G H, 5, 103

H₂X, 183, 195–6

IF, *see* Intermediate frequency amplifier
IFF, *see* Interrogation of Friend or Foe
Imperial War Museum, 17
Ingleby, Peter, 16–17, 20–4, 155
Intermediate frequency amplifier (IF),
    61
Interrogation of Friend or Foe (IFF),
    67
Italy, plans for invasion of, 172

Jackson, Derek, 80–2, 236, 241
Japan, H₂S for offensive against,
    254–6
Jefferson, Sidney, 10
Jelley, J V, 57
Jenkins, J W, 7
Jodrell Bank, vii, viii, 8, 23, 261
Jones, R V, 120, 233, 235–6, 247
Joubert de la Ferté, Air Chief Marshal
    Sir Philip, 45, 48, 54–5, 159
JU88 aircraft, 21, 136, 236

K-band H₂S, 220–3, 240–57
    Mark VI, 224–5
    six-foot scanners, 242–5
Kent, Duke of, 78
Kiel (bombing of), 235
Killip, Flight Lieutenant Len, 186,
    194, 212–3, 140–1, 143, 254
King, Bob, 100
King, Group Captain, 127
Kinsey, Dr Bernard, 92
Kirby, Onslow, 129
Klystron, 31–2, 48, 61–2, 107–8, 139,
    151
    ground trials, Worth Matravers,
        36–43
    and magnetron dispute, 105–7
Königsberg (bombing of), 252

Lancaster bomber, 99, 149, 154, 175,
    177, 184–5, 189, 202, 217,
    225–6, 242–3, 253, 263
Lakenheath (RAF station), 148
Landale, S E A, 53
Langton Matravers, 42, 50

Larnder, H, 15
LCT landing craft, 171–2
Lecher wires, 35
Le Creusot (bombing of), 144
Lee, Sir George (DCD), 28, 39
Leeson House, 42–3, 45–54, 59,
    61–2, 64, 69–70, 73–4, 76, 91,
    101, 103, 106, 122
Leigh, Wing Commander H de V,
    156, 168
Leigh Light Wellington, *see* Wellington
    bomber
Leigh-Mallory, Air Chief Marshal Sir
    Trafford, 241
Leipzig (bombing of), 180–93, 211
Lewis, W B, 5, 10–11, 15, 17, 21, 29,
    37, 45, 49–50, 120, 231, 240
Liberator aircraft, viii, 162–3, 170,
    185, 193, 195
Lincoln aircraft, 202–3
Lindemann, Professor F A, *see*
    Cherwell, Lord
Lion Tamer, 224–5, 230–2, 257
    see also H₂S *and also* Mark VII H₂S
Llewellyn, Colonel, 133
Loch Neagh, 162
Lock–follow AI, 69–84
Lockheed Hudson aircraft, 22
London Aluminium Co, 40
London County Electric Supply Co,
    87, 139
London Midland & Scottish Railway,
    76
Londonderry, Lord, 1–2
Loran, 253
Lovell, Sir Bernard, 18–19, 38–9, 48,
    85, 168, 181, 184, 214, 222, 254
Lovell, Bryan, 116
Lovell, Joyce (*née* Chesterman), 4, 23,
    110–8
Lovell, Susan, 23, 111
Lovett, R A, 195
Luftwaffe, 11, 14–15, 42, 54–6

Magnetic friction clutch, 72–4
Magnetron transmitter, 32–6, 58, 151
    and klystron dispute, 105–7

Magslip, 199
Malvern, 95, 116–7, 120–3, 160, 168,
    254–5
Malvern hills, 207
Manchester University, 70
Mannheim (bombing of), 207
MAPRE, *see* Ministry of Aircraft
    Production Research
    Establishment
Marigold, 53
Marlborough College, 119
Marris, G C, 45
Martini, General Wolfgang, 234
Martlesham Heath (aerodrome), 21, 89
Matthews, L H, 214
McLean, Sir Robert, 134
Mediterranean Allied Air Force, 253
Megaw, E C S, 35–6, 58
Merlin engine, 129–30
'Metadyne' system, 70
Metox receiver, 156–7, 165–6
Metropolitan Vickers, 7, 12, 70, 72,
    74, 76, 78, 107–8, 198
MGB614 motor gunboat, 82–3
'Micropup', *see* VT90
Milch, General Erhard, 233–4
Minelaying, 263–4
Ministry of Aircraft Production, 45,
    87, 121, 133, 140–1, 157–8, 183
Ministry of Aircraft Production
    Research Establishment, 55
Ministry of Defence Air Historical
    Branch, 128
Ministry of Food, 112
Ministry of Information, 111, 113–5
Ministry of Supply, 48–9
Ministry of Works & Buildings, 121
MIT Radiation Laboratory, 145–6,
    247–8
Modulators, 59–61
Monica (tail warning system), 206,
    236, 252
Montgomery, Field Marshal, 242
Mosquito aircraft, 67, 82, 88, 180,
    208, 215, 243–5, 253
'Mouse' stations (Oboe), 88
'Mousetrap', 207–8
Munich Crisis (1938), 3

Nab Tower, 172
Nash & Thompson, 46–7, 56–7, 65, 70, 100, 103, 107, 135, 157, 198
National Physical Laboratory, 2, 5
Navigation and bombing computer, 257–8
Naxos, viii, 165–70, 233–7
NBC, *see* Navigation and bombing computer
Night interception unit, *see* Fighter Interception Unit
North Atlantic, 162
  U-boats and, 155–7
Norway, invasion of, 27
Novorossisk, 151

Oakington (RAF station), 150, 152
Oboe precision bombing, 88, 175–6, 180, 193, 204, 206, 238, 252
Office of Scientific Research & Development, 26
O'Kane, Bernard J, 57, 91–5, 97, 99, 103, 106–7, 126–7, 143, 148, 152, 160, 199, 247–8
Oliphant, Professor M L E, 31–3, 37, 40, 48
Orchis, 53
Ordfordness, 3, 4, 7, 11–12, 61
OSRD, *see* Office of Scientific Research & Development
Overlord, 203
Owen, C S, 53

PAC, *see* Parachute and cable rocket defence weapon
Paraboloids, 39, 41–3, 46–7
Parachute and cable rocket defence weapon, 29
Pathfinder Force, viii, 137–9, 144, 152–3, 160, 162, 173, 175, 178, 181, 183–4, 187, 196–7, 202, 211–2, 243, 251–2
  and H₂S, 148–50
Paul, Commander R T, 172
Paulus, General, 151
PDS, *see* Post Design Services
Pembroke Dock, 22, 24, 27
Perth, 11, 17–20

Peveril Point (Swanage), 53
Philco Corp, 163, 224
Phillipson, J, 56
Physical Society of London, 33
Pic du Midi, 4, 10
Pilsen (bombing of), 207
Pittsburgh, 147
Plan position indicator (PPI), 14, 55–6, 96, 99, 101, 103, 105, 108, 142, 144, 148, 172, 173, 175, 177, 181, 208–9, 217–21
  display, 198–200
Pointblank directive, 194
Poland, invasion of, 8
Pontypool, 95
Poole Harbour, 111
Portal, Air Chief Marshal Sir Charles, 138, 193, 203, 253
Portland, 40
Portsmouth, 171–2
Port Stanley airport, 260
Post Design Services (PDS), 103, 214
PPI, *see* Plan position indicator
Pratt & Whitney engines, 147
Preston Science School, Malvern College, 122
PRF, *see* Pulse repetition frequency
Prince of Wales, 53
Pringle, John, 24, 27
Pulse length, 59–61
Pulse repetition frequency (PRF), 59–61
Purbeck Peninsula, 120
Pye, 23

Quebec Conference, 203

Rabi, I I, 146–7
Radar developments, chronology of, xi–xxi
Radar production unit (RPU), 149
Radar Research & Development Establishment (RRDE), 48
  see also ADEE *and* ADRDE
Radial time base, *see* Plan position indicator
Radio-direction finding, *see* Radar developments, chronology of

Radio Research Board, 27
Radio Research Station, 2
RAE, *see* Royal Aircraft Establishment
RAF, viii, ix, 8, 10, 13–15, 20, 25–6,
    42, 47, 49, 54–6, 58, 66, 82, 88,
    103, 127–32, 139, 194
   Training Command, 12
   see also Bomber Command *and also*
    under specific aircraft
Ramsay, Flight Lieutenant D W C,
    143, 173, 186, 194, 226, 254
Randall, J T, vii, 5, 26, 33, 35, 37, 41
Randall–Boot cavity magnetron, 35, 56
Ratcliffe, J A, 2, 5, 65, 103, 231
Rawnsley, T, 56
RCA, 30
RDF Committee, 11, 101
'Red Neck', 259
Reflex klystron, 61–2
Renwick, Sir Robert, 87, 101, 135,
    140–1, 149, 150, 158, 168, 181,
    183, 198, 201, 208, 213, 215,
    221, 229–30, 242, 245, 247, 255
   biography of, 139–41
Resonant cavities, 31
RF equipment, 218
Rhumbatron, 31–2, 65
Richards, Flying Officer Joe, 143,
    195–6, 254
Rigden, John S, 147
Ringway airport, 149
Ritson, Dr F J U, 70–1, 77, 79, 82–3
Robinson, D M, 48, 145–6, 148,
    162–3, 183–4, 214, 222–3, 230,
    240
Roe, A V, 202, 222
Rolls-Royce, 129–30
Rommel, 133, 151
Roosevelt, President, 26, 132, 156, 172
Roskill, S W, 162, 248–9, 263
Rotterdam, 165, 233
Rotterdam (German code for H2S), 233,
    234
Rowe, A P, 2–3, 5, 10, 12, 15, 18–20,
    27–9, 37, 41–2, 45, 48–9, 53,
    58, 65, 70, 82, 85, 87–8, 91, 94,
    99, 117, 119, 121–3, 167,
    212–3, 230, 231, 241, 247, 252

Royal Aircraft Establishment (RAE),
    15–16, 24–5, 29, 36, 57, 76
Royal Navy, 22
Royal Oak, 17
Royal Society, 5, 11, 17, 24
RPU, *see* Radar production unit
RRDE, *see* Radar Research &
    Development Establishment
RSRE, 128
see also Telecommunication & Research
    Establishment (TRE)
Rudge-Whitworth factory, 134
Ruhr (bombing of), 88, 144, 175
Russia, 151
Rutherford, 5
Ryle, Martin, 127, 239

Salisbury Plain, 93–4
Sansom, Squadron Leader R J, 127
Saundby, Air Marshal Sir Robert, 168,
    181, 208, 215
Saward, Dudley, ix, 17, 87, 101–2,
    106, 108, 140, 175, 178, 181,
    185, 206–8, 229, 236–7, 252,
    282
Sawyer, Lieutenant James H ('Scotty'),
    143–4
Saye, Wing Commander G I L, 102,
    108
Sayers, F, 140–1, 150, 158
Sayers, J, 35, 40, 181
Scan corrector indicator, 200–1, 218–9
Scanner stabilization, 197–8
Scanning tests, 95–6
Scapa Flow, 17
Scarborough, 1–9
Scent Spray (H2S on tank landing
    craft), 172
Schneider works (bombing of), 144
Schenectady–Albany complex, 147
Schnorkel breathing tube, viii, 246–50
Schwenke, Colonel, 233
Scone airport, 10–26, 27–8, 30, 89
Scophony Television Laboratory, 48
SCR584 (US gun-laying radar), 49
SCR720 (US AI radar), 81–2
Sea Lion, 157
Shoenberg, Sir Isaac, 133–4

Short, Lieutenant H F, 171–2
Sicily, invasion of, 172
Sideways-looking Airborne Radar, 259
Siding Spring (N S W), 11
Sinclair, Sir Archibald, 133
Skinner, Herbert W B, 5, 28, 31,
    37–9, 41, 45, 50–2, 56, 61, 63,
    67, 181–2
Skoda armament factory (bombing of),
    207
SLAR, see Sideways-looking Airborne
    Radar
Sleigh, W H, 128–9
Slessor, Air Marshal Sir John (C-in-C
    Coastal Command), 167–70
Smith, Sir Frank, 45–6
    see also Controller of
    Telecommunications Equipment
    (CTE)
Smith, J B, 217–8, 229
Smith, R A, 231
'Soft Sutton tube', 63
Sokol, 157
Southampton, 40, 95, 112
Speer, Albert, 178
Spiral scanner, 57–8
Spitfire aircraft, 78
Split aerial system tests, 95–6
Squegging transmitter, 15
Stalingrad, defence of, 151
Stanford University, 31
Stanmore, 8
Starr, A T, 50, 63, 67, 181
St Athan (aerodrome), 19–27, 53, 155
Staxton Wold, 1–9, 11
Stettin (bombing of), 252
Stirling bomber, 99, 149, 150, 152,
    233–4
    N3724, 182
Strothers, Jane (Mrs S C Curran), 29
Sumburgh Head, 51
'Sunday Soviets', 87–90, 140, 212
Sunderland flying boats, 22, 24, 27
Surbiton, 108
    see also Nash & Thompson
Sutton, R W, 61
Sutton tube, 32

Swanage, 42, 50–1, 53, 69, 77, 103,
    110–8, 126
    Pier, 53
Sylvania, 224

Tait, Air Marshal Sir Victor, 134, 136
Tangmere (aerodrome), 16
Tank landing craft, 171–2
Taylor, Dennis, 171
Tedder, Sir Arthur, 203
Telecommunication Research
    Establishment (TRE), viii, 50,
    55–7, 59, 70–2, 80, 83, 85, 87,
    93, 95, 97, 99–100, 102–3, 106,
    119–20, 122, 128, 134, 141,
    144–7, 157–8, 171, 175, 184,
    202, 211–4, 217, 222–3, 230,
    234, 241, 247, 253, 260
Telefunken Co, 166, 234
Television, 30
Thompson, Dr F C, 158, 186, 254
Thompson, Squadron Leader W H,
    235
Tiger Moth aeroplane, 28
Tirpitz, 137
Titlark, 53, 157
Tizard, H T, 1–3, 11, 16–18, 26, 42,
    58, 104–5, 145
Tizard Committee, 2–3, 11
Tizard Memorial Lecture (1960), 2
Tizard Mission to USA, 147
Tobruk, 132–3
Touch, Gerald, 16–17, 27
TR system, 65, 142–3, 174, 177
    see also Common transmit–receiver
    system
Transmitters, 6–8, 16–17, 58
    Daventry, 3
TRE, see Telecommunication Research
    Establishment
Triode valves, 30–1
Tunis, 172
Turin (bombing of), 153
Tustin, A, 70, 73
Type 182, see Fishpond tests
Type 271 equipment, 53

U-boats, vii, ix, 17, 53, 85, 155–64, 165–70, 246–50
  Anti U-boat Committee, 167
USA, in H₂S, 183–4
USOSRD (US Office of Scientific Research and Development), 26
Usk, 52

V9977, *see* Halifax V9977
Valiant V-bomber, 260
Valves, 25, 30–1
  and CVD (Committee for the Coordination of Valve Development), 31
Varian, R H and S F, 32
VE Day, 254–6
Velodyne, 76
Vickers Valiant V-bomber, 258–60
Victor V-bomber, 258
Vincent, Pilot Officer C E, 127
VJ Day, 254–6
Volga, the river, 136, 150–1
Voss shipyard, 176
VT90, 25, 30, 45
Vulcan V-bomber, 258–60
V-1 flying bombs, 49

Walker, Sergeant, 208
Walther boat, 246, 249
War Cabinet Secretariat, 86
Ward, A G, 41, 48, 50, 63, 67, 181
Warren, Dr J, 92
Wartime civilian life, 110–8
  food queues, 119–25
Watson–Watt, R A, 2–5, 10–12, 21, 24, 26, 28, 133, 168, 183–4, 222–3, 242
Wellington bomber, 79, 81, 156–7, 159–65, 182, 221, 224, 263

Welsh Bicknor, 128
Wembley Laboratory, *see* GEC
Western Electric, 30
Weymouth, 83
Whirligig (H₂S scanner), 222–3, 225–7, 229
Whitaker, A W, 46–7, 100
  see also Nash & Thompson
White, E L C, 15, 71
Whitley bomber, 90, 99
Wilkes, M V, 5
Wilkins, A F, 3–4, 21
Wilkins, M H F, 26
Willett, Captain B R, 51–2
Williams, F C, 70–3, 75–7, 79, 83, 117, 170, 201, 246–7
Wilson, J G, 29
Wimperis, H E, 1–2
'Window', 80–2, 180, 239
Winter Gardens Pavilion, Great Malvern, 123–4
Winward, Squadron Leader W, 81
Wolverhampton, 95
Woolton, Lord, 112
Worcester, 185
Worcester Beacon, 123, 240
World War I, 112
Worth Matravers, 27–46, 51, 55, 59, 61–2, 69, 91, 120, 155
  ground trials at, 36–42
Wyn Williams, C E, 5

X-band, H₂S, 180–93, 220–3

Yagi aerials, 22
Young, Commander R T, 82

Zuckerman, Solly, 104–5